T0192172

Communications in Computer and Information Science   1741

More information about this series at https://link.springer.com/bookseries/7899

Laurence A. F. Park · Heitor Murilo Gomes ·
Maryam Doborjeh · Yee Ling Boo ·
Yun Sing Koh · Yanchang Zhao ·
Graham Williams · Simeon Simoff (Eds.)

Data Mining

20th Australasian Conference, AusDM 2022
Western Sydney, Australia, December 12–15, 2022
Proceedings

 Springer

Editors
Laurence A. F. Park (iD)
Western Sydney University
Sydney, NSW, Australia

Heitor Murilo Gomes (iD)
Victoria University of Wellington
Wellington, New Zealand

Maryam Doborjeh (iD)
Auckland University of Technology
Auckland, New Zealand

Yee Ling Boo (iD)
RMIT University
Melbourne, VIC, Australia

Yun Sing Koh (iD)
University of Auckland
Auckland, New Zealand

Yanchang Zhao (iD)
CSIRO Scientific Computing
Canberra, ACT, Australia

Graham Williams (iD)
Australian National University
Canberra, ACT, Australia

Simeon Simoff (iD)
Western Sydney University
Sydney, NSW, Australia

ISSN 1865-0929   ISSN 1865-0937 (electronic)
Communications in Computer and Information Science
ISBN 978-981-19-8745-8   ISBN 978-981-19-8746-5 (eBook)
https://doi.org/10.1007/978-981-19-8746-5

This Springer imprint is published by the registered company Springer Nature Singapore Pte Ltd.
The registered company address is: 152 Beach Road, #21-01/04 Gateway East, Singapore 189721, Singapore

Preface

It is our great pleasure to present the proceedings of the 20th Australasian Data Mining Conference (AusDM 2022) held at the Western Sydney University, during December 12–15, 2022.

The AusDM conference series first started in 2002 as a workshop initiated by Simeon Simoff (Western Sydney University), Graham Williams (Australian National University), and Markus Hegland (Australian National University). Over the years, AusDM has established itself as the premier Australasian meeting for both practitioners and researchers in the area of data mining (or data analytics or data science). AusDM is devoted to the art and science of intelligent analysis of (usually big) data sets for meaningful (and previously unknown) insights. Since AusDM 2002, the conference series has showcased research in data mining through presentations and discussions on state-of-art research and development. Built on this tradition, AusDM 2022 successfully facilitated the cross-disciplinary exchange of ideas, experiences, and potential research directions, and pushed forward the frontiers of data mining in academia, government, and industry.

To mark its 20th anniversary, the AusDM Festival was organized to include various special sessions such as a Women in Data Mining/Artificial Intelligence Breakfast Panel, a Doctoral Consortium, an Industry and Government Day, Tutorial Sessions, a Student Showcase, etc. There was also a co-located session with the International Federation for Information Processing (IFIP). In addition, a special issue of the Data Science and Engineering journal published by Springer has also been planned.

AusDM 2022 received 44 valid submissions, with authors from 13 different countries. The top five countries, in terms of the number of authors who submitted papers to AusDM 2022, were Australia (66 authors), New Zealand (22 authors), India (17 authors), Poland (11 authors) and the UK (5 authors). All submissions went through a double-blind review process, and each paper received at least three peer reviewed reports. Additional reviewers were considered for a clearer review outcome if review comments from the initial three reviewers were inconclusive.

Out of these 44 submissions, a total of 17 papers were finally accepted for publication. The overall acceptance rate for AusDM 2022 was 38%. Out of the 25 Research Track submissions, nine papers (i.e., 36%) were accepted for publication. Out of the 19 submissions in the Application Track, eight papers (i.e., 42%) were accepted for publication.

The AusDM 2022 Organizing Committee would like to give their special thanks to Fang Chen and Jeffrey Xu Yu for kindly accepting the invitation to give keynote speeches. In addition, the success of various special sessions was attributed to the speakers, namely Ian Oppermann, Stela Solar, Giovanni Russello, Anthony Wong, Rohan Samaraweera, Lin-Yi Chou, and Flora Salim. The committee was grateful to have Fang Chen, Yun Sing Koh, Nandita Sharma, Dilusha Weeraddana, Annelies Tjetjep and Kimberly Beebe as the panellists for the Women in Data Mining/Artificial Intelligence Breakfast Panel. In addition, the committee would like to express their appreciation to Guilio Valentino Dalla Riva and Matthew Skiffington for presenting at the Tutorial Sessions.

The committee would also like to give their sincere thanks to Western Sydney University, for providing admin support and the conference venue, and Springer CCIS and the Editorial Board, for their acceptance to publish AusDM 2022 papers. This will give excellent exposure to the papers accepted for publication. We would also like to give our heartfelt thanks to all student and staff volunteers at the Western Sydney University who did a tremendous job in ensuring a successful 20th anniversary event.

Last but not least, we would like to give our sincere thanks to all delegates for attending the conference this year virtually or in-person at Western Sydney University. We hope that it was a fruitful experience and that you enjoyed AusDM 2022!

December 2022

Laurence A. F. Park
Heitor Murilo Gomes
Maryam Doborjeh
Yee Ling Boo
Yun Sing Koh
Yanchang Zhao
Graham Williams
Simeon Simoff

Organization

General Chairs

Yun Sing Koh   The University of Auckland, New Zealand
Yanchang Zhao   Data61, CSIRO, Australia

Program Chairs (Research Track)

Heitor Murilo Gomes   Victoria University of Wellington, New Zealand
Laurence Park   Western Sydney University, Australia

Program Chair (Application Track)

Maryam Doborjeh   Auckland University of Technology, New Zealand

Program Chair (Industry Track)

Jess Moore   Australian National University, Australia

Special Session Chair

Diana Benavides Prado   The University of Auckland, New Zealand

Publication Chair

Yee Ling Boo   RMIT University, Australia

Diversity, Equity and Inclusion Chair

Richi Nayak   Queensland University of Technology

Finance Chair

Michael Walsh   Western Sydney University

Web Chair

Ben Halstead   The University of Auckland, New Zealand

Publicity Chair

Monica Bian   University of Sydney, Australia

Local Organizing Chairs

Quang Vinh Nguyen   Western Sydney University, Australia
Zhonglin (Jolin) Qu   Western Sydney University, Australia

Tutorial Chair

Varvara Vetrova   University of Canterbury, New Zealand

Doctoral Symposium Chair

Vithya Yogarajan   University of Auckland, New Zealand

Student Showcase Chair

Tony Nolan   G3N1U5 Pty Ltd, Australia

Steering Committee Chairs

Simeon Simoff   Western Sydney University, Australia
Graham Williams   Australian National University, Australia

Steering Committee

Peter Christen   Australian National University, Australia
Ling Chen   University of Technology Sydney, Australia
Zahid Islam   Charles Sturt University, Australia
Paul Kennedy   University of Technology Sydney, Australia
Yun Sing Koh   The University of Auckland, New Zealand
Jiuyong (John) Li   University of South Australia, Australia
Richi Nayak   Queensland University of Technology, Australia
Kok-Leong Ong   RMIT University, Australia
Dharmendra Sharma   University of Canberra, Australia
Glenn Stone   Western Sydney University, Australia
Yanchang Zhao   Data61, CSIRO, Australia

Honorary Advisors

John Roddick   Flinders University, Australia
Geoff Webb   Monash University, Australia

Program Committee

Research Track

Ashad Kabir   Charles Stuart University, Australia
Selasi Kwashie   Data61, CSIRO, Australia
Liwan Liyanage   University of Western Sydney, Australia
Khanh Luong   Queensland University of Technology, Australia
Xueping Peng   University of Technology Sydney, Australia
Weijia Zhang   Southeast University, China
Huma Ameer   National University of Sciences and Technology,
                                  Pakistan
Thirunavukarasu   Queensland University of Technology, Australia
  Balasubramaniam
Guilherme Cassales   University of Waikato, New Zealand
Fabricio Ceschin   Federal University of Parana, Brazil
Rushit Dave   Minnesota State University at Mankato, USA
Nuwan Gunasekara   University of Waikato, New Zealand
Yi Guo   Western Sydney University, Australia
Paul Hurley   Western Sydney University, Australia
Md Zahidul Islam   Khulna University, Bangladesh
Gang Li   Deakin University, Australia
Anton Lord   Leap In!, Australia
Wolfgang Mayer   University of South Australia, Australia
Dang Nguyen   Deakin University, Australia
Hoa Nguyen   Australian National University, Australia
Sharon Torao Pingi   Queensland University of Technology, Australia
Dhananjay Thiruvady   Deakin University, Australia
Russell Thomson   Western Sydney University, Australia
Rosalind Wang   Western Sydney University, Australia
Jie Yang   University of Wollongong, Australia
Vithya Yogarajan   The University of Auckland, New Zealand
Josh Bensemann   The University of Auckland, New Zealand
Evan Crawford   Western Sydney University, Australia
Warwick Graco   Australian Tax Office, Australia
Michal Ptaszynski   Kitami Institute of Technology, Japan
Md Anisur Rahman   Charles Sturt University, Australia

Nicholas Sheppard   Western Sydney University, Australia
Franco Ubaodi   Western Sydney University, Australia

Application Track

Philippe Fournier-Viger   Shenzhen University, China
Boris Bacic   Auckland University of Technology, New Zealand
Xuan-Hong Dang   IBM T. J. Watson Research Center, USA
Zohreh Doborjeh   The University of Auckland, New Zealand
Akbar Ghobakhlou   Auckland University of Technology, New Zealand
Warwick Graco   Australian Tax Office, Australia
Aunsia Khan   National University of Modern Languages,
                                       Pakistan
Edmund Lai   Auckland University of Technology, New Zealand
Jin Li   Data2Action, Australia
Weihua Li   Auckland University of Technology, New Zealand
William Liu   Auckland University of Technology, New Zealand
Jing Ma   Auckland University of Technology, New Zealand
Sam Madanian   Auckland University of Technology, New Zealand
Farhaan Mirza   Auckland University of Technology, New Zealand
Muhammad Marwan Muhammad   Coventry University, UK
  Fuad
Parma Nand   Auckland University of Technology, New Zealand
Minh Nguyen   Auckland University of Technology, New Zealand
Oliver Obst   Western Sydney University, Australia
Yanfeng Shu   CSIRO, Australia
Annette Slunjski   Institute of Analytics Professionals of Australia,
                                       Australia
William Wong   Auckland University of Technology, New Zealand
Yue Xu   Queensland University of Technology, Australia
Sira Yongchareon   Auckland University of Technology, New Zealand
Hamid Abbasi   The University of Auckland, New Zealand
Farnoush Falahatraftar   Polytechnique Montreal, Canada
Alan Litchfield   Auckland University of Technology, New Zealand
Victor Miranda   Auckland University of Technology, New Zealand
Ji Ruan   Auckland University of Technology, New Zealand
Shoba Tegginmath   Auckland University of Technology, New Zealand
Weiqi Yan   Auckland University of Technology, New Zealand

Additional Reviewers

Rasoul Amirzadeh
Fatemah Ansarizadeh

Mike Watts
Zhaofei Wang
Peter Hough

Contents

Research Track

Measuring Content Preservation in Textual Style Transfer 3
Stuart Fitzpatrick, Laurence Park, and Oliver Obst

A Temperature-Modified Dynamic Embedded Topic Model 15
Amit Kumar, Nazanin Esmaili, and Massimo Piccardi

Measuring Difficulty of Learning Using Ensemble Methods 28
Bowen Chen, Yun Sing Koh, and Ben Halstead

Graph Embeddings for Non-IID Data Feature Representation Learning 43
Qiang Sun, Wei Liu, Du Huynh, and Mark Reynolds

Enhancing Understandability of Omics Data with SHAP, Embedding
Projections and Interactive Visualisations 58
*Zhonglin Qu, Yezihalem Tegegne, Simeon J. Simoff, Paul J. Kennedy,
Daniel R. Catchpoole, and Quang Vinh Nguyen*

WinDrift: Early Detection of Concept Drift Using Corresponding
and Hierarchical Time Windows 73
Naureen Naqvi, Sabih Ur Rehman, and Md Zahidul Islam

Investigation of Explainability Techniques for Multimodal Transformers 90
Krithik Ramesh and Yun Sing Koh

Effective Imbalance Learning Utilizing Informative Data 99
Han Tai, Raymond Wong, and Bing Li

Interpretable Decisions Trees via Human-in-the-Loop-Learning 115
Vladimir Estivill-Castro, Eugene Gilmore, and René Hexel

Application Track

A Comparative Look at the Resilience of Discriminative and Generative
Classifiers to Missing Data in Longitudinal Datasets 133
Sharon Torao Pingi, Md Abul Bashar, and Richi Nayak

Hierarchical Topic Model Inference by Community Discovery on Word
Co-occurrence Networks 148
Eric Austin, Amine Trabelsi, Christine Largeron, and Osmar R. Zaïane

UMLS-Based Question-Answering Approach for Automatic Initial Frailty
Assessment 163
  Yashodhya V. Wijesinghe, Yue Xu, Yuefeng Li, and Qing Zhang

Natural Language Query for Technical Knowledge Graph Navigation 176
  Ziyu Zhao, Michael Stewart, Wei Liu, Tim French,
  and Melinda Hodkiewicz

Decomposition of Service Level Encoding for Anomaly Detection 192
  Rob Muspratt and Musa Mammadov

Improving Ads-Profitability Using Traffic-Fingerprints 205
  Adam Gabriel Dobrakowski, Andrzej Pacuk, Piotr Sankowski,
  Marcin Mucha, and Paweł Brach

Attractiveness Analysis for Health Claims on Food Packages 217
  Xiao Li, Huizhi Liang, Chris Ryder, Rodney Jones, and Zehao Liu

SCHEMADB: A Dataset for Structures in Relational Data 233
  Cody Christopher, Kristen Moore, and David Liebowitz

Author Index 245

Research Track

Measuring Content Preservation in Textual Style Transfer

Stuart Fitzpatrick[1]([✉]), Laurence Park[2], and Oliver Obst[2]

[1] School of Computer, Data and Mathematical Sciences, Western Sydney University, Penrith, Australia
S.Fitzpatrick2@westernsydney.edu.au
[2] Centre for Research in Mathematics and Data Science, Western Sydney University, Penrith, Australia
lapark@cdms.westernsydney.edu.au, O.Obst@westernsydney.edu.au

Abstract. Style transfer in text, changing text that is written in a particular style such as the works of Shakespeare to be written in another style, currently relies on taking the cosine similarity of the sentence embeddings of the original and transferred sentence to determine if the content of the sentence, its meaning, hasn't changed. This assumes however that such sentence embeddings are style invariant, which can result in inaccurate measurements of content preservation. To investigate this we compared the average similarity of multiple styles of text from the Corpus of Diverse Styles using a variety of sentence embedding methods and find that those embeddings which are created from aggregated word embeddings are style invariant, but those created by sentence embeddings are not.

Keywords: Style transfer · Content preservation · Cosine similarity

1 Introduction

Making a piece of text that is written in one particular way, its style, be it that of Shakespeare, a tweet or everyday Conversational English and making it appear to be written in another way, such as the works of James Joyce or Mary Shelley, in such as way as to preserve the semantics or meaning of the sentence [6] is the task of textual style transfer.

Style transfer in the domain of natural language processing (NLP) often uses a two-valued metric to measure the success of a model's ability to transfer style: The first of these is Transfer Strength or Transfer Accuracy [5–7], in a nutshell if our task is to transfer from style $\mathbb{A}$ to style $\mathbb{B}$ then the Strength or Accuracy of the style transfer is simply the accuracy at which we successfully transfer the style of the input sentence to our target style. The second of these is Semantic or Content Preservation [5–7], which is put simply, the evaluation of whether the input sentence and the output sentence have the same meaning. For example if Shakespeare is being transferred to Conversational English, then "wherefore art

L. A. F. Park et al. (Eds.): AusDM 2022, CCIS 1741, pp. 3–14, 2022.
https://doi.org/10.1007/978-981-19-8746-5_1

thou Romeo?" should be become "where are you Romeo?" and not just a random sentence from Conversational English (which would be correct style transfer).

Of great difficulty however, is the means by which Semantic Preservation is measured. Previously this has been measured using the Bilingual EvaLuation Understudy (BLEU), however this ultimately proved unfit for the task due to issues such as; the ability to easily manipulate the score of a system [13], discouraging output diversity [15], not upweighting words that are semantically important [14,15] and unreliable correlations between n-gram overlaps and human judgements [2].

To overcome this, an approach of taking the cosine distance (called cosine similarity, in this context) between the sentence embeddings of the input and output sentences [15] was developed. The specific methods that sentences are embedded with vary, but are typically aggregated word embeddings, where the individual words are embedded using word2vec or GloVe [3,4,15].

In previous work, when evaluation of content preservation is desired [3,4,6, 15], each pair of input and output sentences has been mapped to vectors using word embedding methods and then the similarity of the sentences is compared; if the vectors are similar, then it is assumed that the content/meaning of the two sentences is the same.

However, if a style transfer system incorrectly outputs a sentence of the same style as the input, then the sentences will be incorrectly evaluated as being more similar than they are intended to be, and conversely, the nature of a style transfer system requires that the input and output sentences are of different styles, does that not mean then that correct style transfer is penalised when it comes to content preservation? So there is clearly an assumption that sentence embeddings are style invariant.

In this paper, we investigate the assumption that sentence embeddings are inherently style invariant. We make as a research contribution; showing that aggregated word embeddings, but not pre-trained sentence, embeddings are indeed style invariant.

The remainder of this paper is organised as follows; in Sect. 2 we discuss related previous work on content preservation for style transfer, in Sect. 3 we detail our investigation into the style invariance of sentence embeddings, in Sect. 4 we present and discuss our results, and we conclude the paper in Sect. 5.

2   Background and Motivation

There has been some work into finding a method by which content preservation within style transfer (and other related areas such as neural machine translation and semantic similarity) may be measured, namely the use of the cosine similarity on the input and output sentences [15]. In this section we discuss cosine similarity and its use, the disentanglement of style and content and the style invariance assumption of sentence embeddings.

2.1   Cosine Similarity

Let us consider two embedded sentences with respect to a style transfer system, an input $\mathbf{A}$ and an output sentence $\mathbf{B}$. The cosine similarity (distance) between the two sentences is given by:

$$\text{Cossim}(\mathbf{A}, \mathbf{B}) = \cos(\theta_{\mathbf{A},\mathbf{B}}) = \frac{\langle \mathbf{A}, \mathbf{B} \rangle}{\|\mathbf{A}\| \cdot \|\mathbf{B}\|} \quad \mathbf{A}, \mathbf{B} \in \mathbb{R}^n \qquad (1)$$

where, $\langle \mathbf{A}, \mathbf{B} \rangle$ is the inner product of $\mathbf{A}$ and $\mathbf{B}$, $\|\mathbf{A}\|$ and $\|\mathbf{B}\|$ denotes the norms of $\mathbf{A}$ and $\mathbf{B}$ respectively, $\mathbb{R}$ is the set of real numbers and n is the predefined size of the embedding space.

Recall that a similarity of 0 means that there is no similarity between the two sentence embeddings, a similarity of 1 means the sentence embeddings are exactly the same (to a scaling value) and a similarity -1 means the sentence embeddings are exactly opposite (to a scaling value). Two sentences of different styles, such as the input and output of a style transfer system, carry approximately the same meaning if their cosine similarity is sufficiently close to 1.

In [3] and [4] cosine similarity is used by creating three different aggregated word embeddings, element-wise maximum, element-wise minimum and element-wise mean, then concatenating these into a single vector which has its similarity measured. In the case of [3] the GloVe-wiki-gigaword-100 [9] embeddings are used and in the case of [4] the word2vec [8] embeddings are used.

The work of [15], where the use of cosine similarity in this way was popularised, the cosine similarity is directly used on the sentences that have been embedded with the SentPiece[1] embedder. The work of [6] takes the same approach as [15], with the caveats that the sentence embeddings are taken from the [CLS] vectors of a RoBERTa style classifier that they fine-tuned and that the similarity is later incorporated into an overall style transfer metric.

2.2   Disentanglement of Style and Content

Due to the discreteness of natural language, the fact that short sentences do not contain much style information and a severe lack of parallel corpora, it is not possible at present to fully disentangle the style and content of text, like what is possible in image processing [3,11,12]. Full disentanglement, however, may not be necessary. In the event that style and content cannot be fully disentangled, it should still be possible to attain a partial separation of style and content, such that it is possible to determine the similarity of content between two (or more) sentences.

It is to that end that much previous work has separated the measurement of correct style transfer and preservation of content, with much work going into the latter such as in [7] and [5]. The method of choice in many papers for measuring the content preservation between an input sentence and its transferred

---

[1] https://github.com/google/sentencepiece.

counterpart, likely chosen for its simplicity, is as previously mentioned, cosine similarity.

However, with the exception of the first authors to replace BLEU with cosine similarity, [15], which is a language translation task (so the overlap isn't perfect), not a style transfer task, no implementation of cosine similarity as the content preservation metric has, to the authors' knowledge, actually verified that the underlying sentence representations (at least partially) disentangle style and content and thus that, cosine similarity does what is intended.

2.3   The Style Invariant Embedding Assumption

More recent work, that which uses cosine similarity over older metrics such as BLEU [3,4,6,15], do not seem to make a mention of the assumption of their work relies upon, namely the one mentioned in Sect. 1:

> *The process of generating a sentence embedding implicitly removes non-semantic (i.e. stylistic) information, that is, it is style invariant.*

The reason it is important that the sentence embeddings are style invariant, is due to the following situations: (1) If the task is to transfer from style $\mathbb{A}$ to $\mathbb{B}$ and the style transfer system incorrectly outputs a sentence in style $\mathbb{A}$, then if the sentence embeddings are not style invariant, then the input/output pair of sentences will be evaluated as more similar to each other than is desired (especially in the case where style $\mathbb{A}$ has had a sentence output at random). (2) If the task is to transfer from style $\mathbb{A}$ to $\mathbb{B}$ and the style transfer system correctly outputs to a sentence in style $\mathbb{B}$, then as each style contains different sets of words that have the same meaning, the input/output pair of sentences will be evaluated as being less similar than is desired.

Fortunately, investigating whether the various sentence embedding methods are style invariant is rather straightforward. Let us assume for a moment that it is the case that the assumption is true. In a corpus of non-parallel sentences in multiple styles, then it should be the case that any given style within the corpus has an equal probability of being most similar to any other style, that it is equivalent to a single sample from a discrete uniform distribution over the number of styles (11 in the case of our data). If however the embedding methods are not style invariant then it is the case that multiple styles will demonstrate self-similarity (being most similar to itself).

3   Experiment

The purpose of this research is to determine if the various sentence embedding approaches (aggregated word embeddings and pre-trained sentence embedders) are style invariant. In this section, we evaluate the style invariance of three different sentence embedding approaches; two aggregated word embeddings and one pre-trained sentence embedding model, by computing the average cosine similarity of each of eleven styles of the Corpus of Diverse Styles (CDS), and observing how often a style demonstrates self-similarity (that is, it is most similar to itself). If self-similarity is rare or non-existent when using a specific embedding method, then it holds that the sentence embedding method is style-invariant. We then use `t-SNE` to dimensionally reduce the sentence embeddings to determine if there is any clustering of the embeddings with respect to style of the sentence it represents.

It should be noted that this task is a challenge to solve without access to a corpus of sentences in different styles that are parallel with respect to content (i.e. they have the same content in a variety of styles.). The use of the CDS, which contains non-parallel sentences in multiple different styles will aid with task by providing sentences that can only be similar due to the style of the sentence and the use of common words across many or all styles (such as *the, and, of,* and *by*), and thus if a style shows self-similarity it must be due to one (or both) of those properties.

3.1 Dataset

The Corpus of Diverse Styles (CDS) [6] is a dataset that consists of non-parallel sentences in the following eleven styles, 1000 sentences each, collated from a variety of sources:

1. 1810–1820
2. 1890–1900
3. 1990–2000
4. African American English Tweets (AAE)
5. King James Bible
6. English Tweets
7. James Joyce
8. Song Lyrics
9. Romantic Poetry
10. Shakespeare
11. Switchboard

The data also contains generated sentences from the Style TRansfer as Paraphrasing (STRAP) of [6], but these are not used to ensure that our results are not biased by this model.

3.2 Procedure

The aim of the this experiment is to determine the style invariance of various sentence embedding techniques and therefore the appropriateness of each of these methods for determining the content preservation of various style transfer systems. To achieve this we undertake the following procedure. Each of the sentences in each of the styles were embedded using each of the following methods.

– `GloVe-wiki-gigaword-300` [9] concatenated with element-wise summation.

- `FastTest-Crawl-300d-2M` [1] concatenated with element-wise summation.
- `Distilroberta-base` [10]

In the cases where the embedding method is a sentence embedding model, no further action is undertaken, in the cases where the embedding method is a word embedding model, then a sentence embedding is generated by taking the element-wise sum of each of the embedded words, to create a sentence embedding with a dimensionality identical to that of the word embeddings.

For each of these sentence embeddings, the cosine similarity of that sentence embedding versus every sentence embedding in every style (including its own style) are calculated. Each sentence embedding in a style, which has associated similarities for each other sentence embedding it was compared with in each given style, has the mean of these similarities taken. These means, when taken together comprise the *sampling distribution* of style **A** when compared to style **B**.

Next, the mean of each of these sampling distributions is taken as the estimate of the input style's similarity to the target style. The tables that show all 121 pairwise comparisons for each of the three sentence embedding methods, is shown below in Sect. 4. Finally, for each style, we find which style it is the most similar to (i.e. has the highest average cosine similarity). This will allow us to determine if the embeddings are style invariant, by showing how close each of the style to style similarity comparison is to the result of pure chance.

4   Results and Discussion

Provided in Tables 1, 2 and 3 are the style similarity estimates for the eleven different styles in the CDS, as labelled in Sect. 3.1, using each of the different embedding methods, with the highest similarity in bold.

Table 1. Average Cosine Similarity of Each of the Eleven Styles in the CDS as labelled in Sect. 3.1 - `GloVe-wiki-gigaword-300`, bold indicates the highest level of similarity.

	1	2	3	4	5	6	7	8	9	10	11
1	**0.775**	0.765	0.730	0.558	0.768	0.591	0.588	0.653	0.572	0.654	0.758
2	**0.757**	0.749	0.718	0.550	0.750	0.581	0.579	0.643	0.561	0.641	0.745
3	**0.732**	0.724	0.696	0.541	0.720	0.568	0.558	0.629	0.542	0.622	0.727
4	0.594	0.591	0.569	0.521	0.594	0.512	0.466	0.575	0.456	0.563	**0.656**
5	0.766	0.753	0.717	0.554	**0.782**	0.582	0.583	0.647	0.572	0.657	0.740
6	0.624	0.617	0.597	0.511	0.620	0.520	0.485	0.579	0.471	0.570	**0.666**
7	**0.606**	0.601	0.576	0.452	0.602	0.475	0.471	0.527	0.456	0.525	0.605
8	0.706	0.699	0.673	0.570	0.702	0.578	0.547	0.646	0.533	0.638	**0.743**
9	0.601	0.592	0.565	0.448	**0.607**	0.467	0.464	0.522	0.464	0.529	0.592
10	0.691	0.681	0.650	0.545	0.696	0.559	0.533	0.622	0.523	0.632	**0.713**
11	0.756	0.748	0.718	0.606	0.743	0.617	0.581	0.687	0.562	0.676	**0.806**

Table 2. Average Cosine Similarity of Each of the Eleven Styles in the CDS as labelled in Sect. 3.1 - `FastText-Crawl-300d-2M`, bold indicates the highest level of similarity.

	1	2	3	4	5	6	7	8	9	10	11
1	0.591	0.580	0.559	0.463	0.617	0.464	0.399	0.550	0.470	0.535	**0.624**
2	0.569	0.564	0.546	0.461	0.590	0.457	0.389	0.540	0.455	0.517	**0.612**
3	0.552	0.549	0.537	0.457	0.569	0.449	0.380	0.530	0.442	0.500	**0.598**
4	0.466	0.472	0.466	0.478	0.493	0.440	0.338	0.510	0.390	0.473	**0.573**
5	0.586	0.572	0.549	0.466	**0.656**	0.460	0.400	0.549	0.481	0.549	0.612
6	0.472	0.472	0.464	0.448	0.492	0.431	0.334	0.496	0.387	0.465	**0.561**
7	0.409	0.409	0.396	0.355	0.436	0.344	0.296	0.400	0.341	0.391	**0.448**
8	0.508	0.506	0.495	0.467	0.534	0.444	0.355	0.532	0.411	0.492	**0.603**
9	0.490	0.483	0.468	0.398	**0.528**	0.392	0.339	0.464	0.415	0.461	0.510
10	0.497	0.490	0.473	0.439	0.544	0.421	0.347	0.498	0.410	0.501	**0.560**
11	0.560	0.558	0.543	0.506	0.577	0.490	0.390	0.581	0.444	0.535	**0.685**

Table 3. Average Cosine Similarity of Each of the Eleven Styles in the CDS as labelled in Sect. 3.1 - `Distilroberta-base`, bold indicates the highest level of similarity.

	1	2	3	4	5	6	7	8	9	10	11
1	**0.972**	**0.972**	0.968	0.967	0.960	0.970	0.966	0.966	0.963	0.964	0.963
2	0.972	**0.973**	0.970	0.966	0.956	0.969	0.964	0.966	0.960	0.962	0.960
3	0.968	0.970	**0.976**	0.964	0.953	0.964	0.960	0.965	0.957	0.958	0.955
4	0.967	0.965	0.964	**0.977**	0.969	0.965	0.972	0.973	0.969	0.966	0.969
5	0.960	0.957	0.955	0.970	**0.972**	0.961	0.967	0.968	0.970	0.970	0.970
6	**0.970**	0.969	0.964	0.966	0.961	**0.970**	0.965	0.964	0.963	0.964	0.964
7	0.968	0.966	0.963	**0.973**	0.967	0.967	0.970	0.969	0.967	0.965	0.968
8	0.966	0.966	0.966	**0.973**	0.967	0.964	0.969	0.972	0.968	0.965	0.967
9	0.962	0.959	0.957	0.970	0.969	0.962	0.967	0.968	**0.971**	0.969	0.968
10	0.965	0.962	0.960	0.966	0.967	0.965	0.964	0.965	0.968	**0.973**	0.967
11	0.963	0.960	0.957	**0.970**	**0.970**	0.964	0.968	0.968	0.970	0.969	0.970

Provided in Tables 4, 5 and 6 are each style (as labelled in Sect. 3.1), and which style it is most similar to. Any self similarity (a style being most similar to itself) is shown in bold. It is interesting to note that only `Distilroberta-base` had any occurrences of tied greatest similarity.

Table 4. Each Style and its Most Similar Style as labelled in Sect. 3.1 - `GloVe-wiki-gigaword-300`, bold indicates self-similarity.

Style	1	2	3	4	5	6	7	8	9	10	11
Is most similar to	**1**	1	1	11	**5**	11	1	11	5	11	**11**

Table 5. Each Style and its Most Similar Style as labelled in Sect. 3.1 - `FastTest-Crawl-300d-2M`, bold indicates self-similarity.

Style	1	2	3	4	5	6	7	8	9	10	11
Is most similar to	11	11	11	11	**5**	11	11	11	5	11	**11**

Table 6. Each Style and its Most Similar Style as labelled in Sect. 3.1 - `Distilroberta-base`, bold indicates self-similarity.

Style	1	2	3	4	5	6	7	8	9	10	11
Is most similar to	1, 2	**2**	**3**	**4**	**5**	1, 6	4	4	9	10	4, 5

In Table 7, it is shown how many times each embedding method results in self similarity.

Table 7. Self Similarity Counts and Proportions for Each of the Three Embedding Methods

	GloVe-wiki-gigaword-300	FastTest-Crawl-300d-2M	Distilroberta-base
Count	3	2	6
Percentage	27.3	18.2	54.5

As there are eleven styles in the CDS, if the various sentence embeddings were indeed fully style invariant, then each style has an equal probability of being as similar to itself as any other style (i.e. the style similarity is discretely uniform over the number of styles), thus it would not be unreasonable to see approximately $\frac{1}{11} \approx 9\%$ occurrence of self-similarity. Although, as stated in Sect. 2.2, style and content are (at least at present) not fully separable from one another, 9% provides a maximum lower bound that sentence embedding methods can strive towards.

Table 7 shows that `Distilroberta-base` performs quite poorly with over half of the styles as embedded showing self-similarity. `GloVe-wiki-gigaword-300` does much better with only three styles showing self-similarity and `FastTest-Crawl-300d-2M` does better again with only two styles showing self-similarity.

From these results it can be concluded that `Distilroberta-base` performs quite poorly at being style invariant, with `GloVe-wiki-gigaword-300` performing twice as well and `FastTest-Crawl-300d-2M` performing the best with only two occurrences of self-similarity and is the closest to the target of 9% ($\frac{1}{11}$).

Shown in the Figs. 1, 2 and 3 are the `t-SNE` plots for the sentence embeddings normalised to length 1.

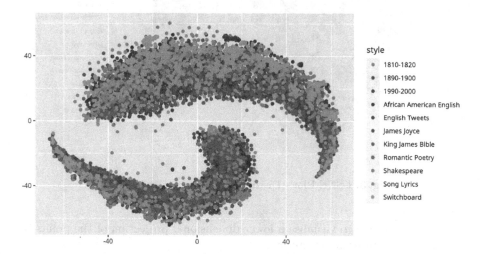

Fig. 1. t-SNE Plot - GloVe aggregated word embeddings

As can be seen in Table 7, both types of aggregated word embeddings have only 3 and 2 cases of self-similarity (GloVe and FastText, respectively) compared to the comparatively high 6 in the case of the dedicated sentence embedding model.

Given that some of the similarities are quite close (as seen in Tables 1, 2 and 3), it is not unreasonable to believe that at least some of these are due to chance.

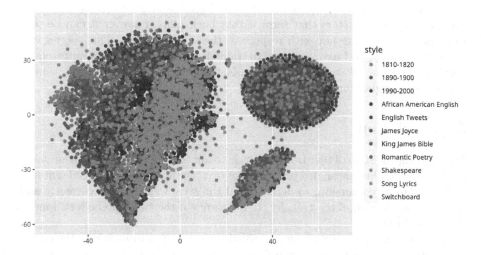

Fig. 2. t-SNE Plot - Fast text aggregated word embeddings

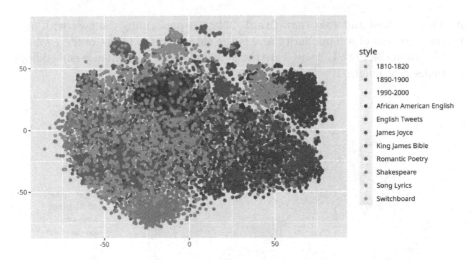

Fig. 3. t-SNE Plot - Distilroberta sentence embeddings

It is hence desirable to visualise a lower dimensional projection of the points to see if there is indeed any clustering.

Figure 1, the t-SNE plot for the aggregated GloVe word embeddings, shows a classic 'Swiss Roll' shape. Figure 2, the corresponding plot for FastText, shows three distinct groups of points. Figure 3, the corresponding plot for Distilroberta shows a relatively even spread of points, with one style clustering on the right hand side.

First, looking at the case of Distilroberta, it is clear to see that, although there is a good deal of overlap between each of the styles, several styles, most notably 1990–2000, do have a relatively clear, well-defined centre, so although some level of style disentanglement has been achieved, ideally something much better can be done. Next, looking at both Figs. 1 and 2, it can be seen that there are distinct clusters that form within both plots, however it can be seen that these clusters are not with respect to the sentence style. Thus aggregated word embeddings (at least compared to trained sentence embedding approaches), particularly those of FastTest-Crawl-300d-2M can be seen as having a much better disentanglement of style.

5   Conclusion

Transferring the style of text from one style to another, requires the measurement of two separate, but entangled properties of text; the style of the original sentence and its transferred version, to determine if the style was changed correctly and content preservation of the transfer, to determine if the two sentences still mean the same thing.

The measurement of content preservation currently utilises taking the cosine similarity of the sentence embeddings of the input and output sentences of the style transfer system. This however relies on the assumption that sentence embedding methods are style invariant.

In this work, using previously existing data, we estimated the average similarity of multiple different textual styles to each other when embedded using various sentence embedding methods and found that sentence embeddings made from aggregated word embeddings have acceptable levels of style invariance, but that pre-trained sentence embeddings do not. We further also generated and plotted, dimensionally reduced points to investigate the nature of any of any clustering.

Possible future work could be undertaken along two paths. First is to investigate why dedicated sentence embedding models such as `Distilroberta-base` are worse at removing the stylistic properties than aggregated word embeddings. Second is to investigate if it is possible to explicitly train a sentence embedding model to remove as much style from a sentence as possible.

References

1. Bojanowski, P., Grave, E., Joulin, A., Mikolov, T.: Enriching word vectors with subword information. Trans. Assoc. Comput. Linguist. **5**, 135–146 (2017)
2. Callison-Burch, C., Osborne, M., Koehn, P.: Re-evaluating the role of bleu in machine translation research. In: 11th Conference of the European Chapter of the Association for Computational Linguistics, pp. 249–256 (2006)
3. Fu, Z., Tan, X., Peng, N., Zhao, D., Yan, R.: Style transfer in text: exploration and evaluation. arXiv preprint arXiv:1711.06861 (2017)
4. Gong, H., Bhat, S., Wu, L., Xiong, J., Hwu, W.: Reinforcement learning based text style transfer without parallel training corpus. arXiv preprint arXiv:1903.10671 (2019)
5. Jin, D., Jin, Z., Hu, Z., Vechtomova, O., Mihalcea, R.: Deep learning for text style transfer: a survey. Comput. Linguist. **48**(1), 155–205 (2022)
6. Krishna, K., Wieting, J., Iyyer, M.: Reformulating unsupervised style transfer as paraphrase generation. arXiv preprint arXiv:2010.05700 (2020)
7. Li, J., Jia, R., He, H., Liang, P.: Delete, retrieve, generate: a simple approach to sentiment and style transfer. In: Proceedings of the 2018 Conference of the North American Chapter of the Association for Computational Linguistics: Human Language Technologies, New Orleans, Louisiana, Volume 1 (Long Papers), pp. 1865–1874. Association for Computational Linguistics (2018). https://doi.org/10.18653/v1/N18-1169. https://www.aclweb.org/anthology/N18-1169
8. Mikolov, T., Sutskever, I., Chen, K., Corrado, G.S., Dean, J.: Distributed representations of words and phrases and their compositionality. In: Advances in Neural Information Processing Systems, vol. 26 (2013)
9. Pennington, J., Socher, R., Manning, C.D.: Glove: global vectors for word representation. In: Proceedings of the 2014 Conference on Empirical Methods in Natural Language Processing (EMNLP), pp. 1532–1543 (2014)
10. Sanh, V., Debut, L., Chaumond, J., Wolf, T.: DistilBERT, a distilled version of BERT: smaller, faster, cheaper and lighter. arXiv abs/1910.01108 (2019)
11. Shen, T., Lei, T., Barzilay, R., Jaakkola, T.: Style transfer from non-parallel text by cross-alignment. In: Advances in Neural Information Processing Systems, pp. 6830–6841 (2017)

12. Subramanian, S., Lample, G., Smith, E.M., Denoyer, L., Ranzato, M., Boureau, Y.L.: Multiple-attribute text style transfer. arXiv preprint arXiv:1811.00552 (2018)
13. Tikhonov, A., Shibaev, V., Nagaev, A., Nugmanova, A., Yamshchikov, I.P.: Style transfer for texts: retrain, report errors, compare with rewrites. arXiv preprint arXiv:1908.06809 (2019)
14. Wang, A., Cho, K., Lewis, M.: Asking and answering questions to evaluate the factual consistency of summaries. arXiv preprint arXiv:2004.04228 (2020)
15. Wieting, J., Berg-Kirkpatrick, T., Gimpel, K., Neubig, G.: Beyond bleu: training neural machine translation with semantic similarity. arXiv preprint arXiv:1909.06694 (2019)

A Temperature-Modified Dynamic Embedded Topic Model

Amit Kumar[1,2]([✉]) [iD], Nazanin Esmaili[1] [iD], and Massimo Piccardi[1] [iD]

[1] University of Technology Sydney, Broadway, Sydney, NSW 2007, Australia
amit270980@gmail.com, {nazanin.esmaili,massimo.piccardi}@uts.edu.au
[2] Food Agility CRC Ltd., Pitt St., Sydney, NSW 2000, Australia

Abstract. Topic models are natural language processing models that can parse large collections of documents and automatically discover their main topics. However, conventional topic models fail to capture how such topics change as the collections evolve. To amend this, various researchers have proposed dynamic versions which are able to extract sequences of topics from timestamped document collections. Moreover, a recently-proposed model, the dynamic embedded topic model (DETM), joins such a dynamic analysis with the representational power of word and topic embeddings. In this paper, we propose modifying its word probabilities with a temperature parameter that controls the smoothness/sharpness trade-off of the distributions in an attempt to increase the coherence of the extracted topics. Experimental results over a selection of the COVID-19 Open Research Dataset (CORD-19), the United Nations General Debate Corpus, and the ACL Title and Abstract dataset show that the proposed model – nicknamed DETM-tau after the temperature parameter – has been able to improve the model's perplexity and topic coherence for all datasets.

Keywords: Topic models · Neural topic models · Dynamic topic models · Dynamic embedded topic model · Deep neural networks

1 Introduction

Topic models are natural language processing (NLP) models which are able to extract the main topics from a given, usually large, collection of documents. In addition, topic models are able to identify the proportions of the topics in each of the individual documents in the given collection, which can be useful for their categorization and organization. As a machine learning approach, topic models are completely unsupervised and, as such, they have proved a very useful tool for the analysis of large amounts of unstructured textual data which would be impossible to tackle otherwise. Thanks to their flexibility and ease of use, topic

Supported by funding from Food Agility CRC Ltd, funded under the Commonwealth Government CRC Program. The CRC Program supports industry-led collaborations between industry, researchers and the community.

models have found application in domains as diverse as finance [8, 18], news [25], agriculture [9], social media [1, 18], healthcare [2, 22, 27] and many others.

Among the topic models proposed to date, latent Dirichlet allocation (LDA) [7] is broadly regarded as the most popular. Its simple, fundamental assumption is that every word in each document of the given collection is associated with a specific "topic". In turn, a topic is represented simply as a dedicated probability distribution over the words in the given vocabulary. Completed by a Dirichlet prior assumption over the topic proportions of each document, LDA has proved at the same time accurate and efficient. However, conventional topic models such as LDA are unable to analyse the sequential evolution of the topics over different time frames. This could be important, instead, for collections that exhibit substantial evolution over time. For instance, a collection of COVID-19-related articles may predominantly display topics such as "outbreak" and "patient zero" in its early stages, "lockdowns" and "vaccine development" in later stages, and "vaccination rates" and "boosters" in the present day.

To analyze the topics over time, one could in principle just partition the document collection into adequate "time slices" (e.g., months or years), and apply a conventional topic model separately over each time slice. However, this would fail to capture the continuity and the smooth transitions of the topics over time. For this reason, Lafferty and Blei in [13] have proposed a *dynamic topic model* (DTM) which is able to extract the topics from each time slice while taking into account the topics' continuity and temporal dynamics. Motivated by the representational power of word embeddings in NLP, Diang et al. in [10] have recently proposed a *dynamic embedded topic model* (DETM) which integrates DTM with embedded word representations. Since word embeddings can be pre-trained in a completely unsupervised way over large amounts of text, an embedded model such as DETM can take advantage of the information captured by the word embeddings' pre-training.

However, a common limitation for all these topic models is that they cannot be easily tuned to explore improvements of the performance evaluation measures. For this reason, in this paper we propose adding a tunable parameter (a "temperature") to the word distributions of DETM to attempt increasing the model's performance. We have tested the proposed model, aptly nicknamed *DETM-tau* after the temperature parameter, over three diverse and probing datasets: a time-sliced subset of the COVID-19 Open Research Dataset (CORD-19) [24], the United Nations (UN) General Debate Corpus [3], and the ACL Title and Abstract Dataset [5], comparing it with the best dynamic topic models from the literature such as DTM and DETM. The experimental results show that the proposed model has been able to achieve higher topic coherence and also lower test-set perplexity than both DTM and DETM in all cases.

The rest of this paper is organized as follows: the related work is presented in Sect. 2, including a concise review of the key topic models. DETM is recapped in Sect. 3.1, while the proposed approach is presented in Sect. 3.2. The experiments and their results are presented in Sect. 4. Eventually, the conclusion is given in Sect. 5.

2   Related Work

In this section, we review the topic models that are closely related to the proposed work, such as latent Dirichlet allocation (LDA), dynamic topic models, and topic models based on word and topic embeddings.

Let us consider a document collection, D, with an overall vocabulary containing V distinct words. In LDA, the generic n-th word in the d-th document can be noted as $w_{d,n}$, and simply treated as a categorical variable taking values in index set $[1 \ldots V]$. One of the key assumptions of LDA is that each such word is uniquely assigned to a corresponding *topic*, $z_{d,n}$, which is another categorical variable taking values in set $[1 \ldots K]$, where K is the number of topics that we choose to extract from the collection. In turn, each topic has an associated probability distribution over the words in the vocabulary, $\beta_k, k = 1 \ldots K$, which accounts for the word frequencies typical of that specific topic. The full model of LDA can be precisely formulated and understood in terms of the following *generative model*, which is a model able to generate "synthetic" documents by orderly sampling from all the relevant distributions:

- For the d-th document, draw a K-dimensional vector, θ_d, with its topic proportions:
    - $\theta_d \sim \mathrm{Dir}(\theta_d|\alpha)$
- For each word in the d-th document:
    - Draw its topic: $z_{d,n} \sim \mathrm{Cat}(\theta_d)$
    - Draw the word from the topic's word distribution:
    - $w_{d,n} \sim \mathrm{Cat}(\beta_{z_{d,n}})$.

In the above model, the first step for each document is to sample its topic proportions, θ_d, from a suitable Dirichlet distribution, $\mathrm{Dir}(\theta_d|\alpha)$. Once the topic proportions are given, the next step is to sample all of the document's words, by first sampling a topic, $z_{d,n}$, from categorical distribution[1] $\mathrm{Cat}(\theta_d)$, and then sampling the corresponding word, $w_{d,n}$, from the word distribution indexed by $z_{d,n}$, $\mathrm{Cat}(\beta_{z_{d,n}})$.

Overall, LDA is a computationally-efficient model that can be used to accurately extract the topics of a given training set of documents, and simultaneously identify the topic proportions of each individual document. LDA can also be applied to a given *test set*; in this case, the parameters of the Dirichlet distribution, α, and the word distributions, β, are kept unchanged, and only the topic proportions for the given test documents are inferred. LDA has also spawned a large number of extensions and variants, including hierarchical versions [12,16], sequential versions [21], class-supervised versions [21], sparse versions [19,26,28], and many others. However, the extensions that are closely relevant to our work are the dynamic topic model (DTM) [13], the embedded topic model (ETM) [11], and the dynamic embedded topic model (DETM) [10]. We briefly review DTM and ETM hereafter, while we recap DETM in greater detail in Sect. 3.

---

[1] Otherwise known as the multinomial distribution. The recent literature on variational inference seems to prefer the "categorical distribution" diction.

DTM is a topic model that captures the evolution of the topics in a corpus of documents that is sequentially organized (typically, along the time dimension). The corpus is first divided up into "time slices" (i.e., all the documents sharing the same time slot), and then the topics are extracted from each slice taking into account a dynamic assumption. For reasons of inference efficiency, DTM uses a logistic normal distribution, $\mathcal{LN}(\theta|\alpha)$, instead of a Dirichlet distribution to model the topic proportions of the individual documents. In addition, the samples of the logistic normal distribution are obtained by explicitly sampling a Gaussian distribution of equivalent parameters, and then applying the softmax operator, $\sigma(\cdot)$, to the Gaussian samples. The sequential dependencies between the time slices are captured by a simple dynamical model:

$$\alpha^t \sim \mathcal{N}(\alpha^{t-1}, \delta^2 I)$$
$$\beta^t \sim \mathcal{N}(\beta^{t-1}, \sigma^2 I)$$

(1)

where α^t are the parameters of the logistic normal distribution over the topics at time t, and β^t is the matrix of all the word distributions (in logit scale), also at time t. The rest of the generative model for slice t can be expressed as:

- For the d-th document, draw its topic proportions (logit scale):
    - $\theta_d \sim \mathcal{N}(\alpha^t, a^2 I)$.
- For each word in the d-th document:
    - Draw its topic: $z_{d,n} \sim \mathrm{Cat}(\sigma(\theta_d))$
    - Draw the word from the topic's word distribution:
    - $w_{d,n} \sim \mathrm{Cat}(\sigma(\beta_{z_{d,n}}))$.

DTM has proved capable of good empirical performance, and its inference is provided by efficient variational methods [13]. However, both LDA and DTM might lead to poor modelling in the presence of very large vocabularies, especially if the corpus is not sufficiently large to allow accurate estimation of the word probabilities. A possible mollification consists of substantially pruning the vocabulary, typically by excluding the most common and least common words. However, this carries the risk of excluding important terms a priori. The embedded topic model (ETM) [11] aims to overcome the limitations of categorical word distributions such as those of LDA and DTM by leveraging *word embeddings* [4,15].

In ETM, each distinct word in the vocabulary is represented as a point in a standard word embedding space (typically, 300-1024D). Each topic, too, is represented as a point (a sort of "average") in the same embedding space. The compatibility between a word and a topic is then simply assessed by their dot product, and the probability of the word given the topic is expressed as in a common logistic regression classifier. The full generative model of ETM can be given as:

- For the d-th document, draw its topic proportions (logit scale):
    - $\theta_d \sim \mathcal{N}(0, I)$

- For each word in the d-th document:
  - Draw its topic: $z_{d,n} \sim \mathrm{Cat}(\sigma(\theta_d))$
  - Draw the word from the topic's word distribution:
  - $w_{d,n} \sim \mathrm{Cat}(\sigma(\rho^\top \eta_{z_{d,n}}))$

In the above, we have noted as ρ the word embedding matrix, which contains the embeddings of all the words in the given vocabulary. Assuming a dimensionality of L for the embedding space, ρ's size is $L \times V$. In turn, with notation η_k we have noted the embedding of the k-th topic. Therefore, the dot product $\rho^\top \eta_k$ evaluates to a V-dimensional vector which, suitably normalised by the softmax, returns the probabilities for the word distribution of topic k.

The ETM is a powerful topic model that joins the advantages of LDA with the well-established word embeddings. The main benefit brought by the word embeddings is that they can be robustly pre-trained using large amounts of unsupervised text from a relevant domain (potentially, even the collection itself). During training of the ETM, a user can choose to either 1) use the pre-trained word embeddings, keeping them fixed, or 2) load them as initial values, but update them during training. In alternative, a user can also choose to update the word embeddings during training, but initialise them from arbitrary or random values (in this case, not taking advantage of pre-training). Dieng *et al.* in [11] have shown that the ETM has been able to achieve higher topic coherence and diversity than LDA and other contemporary models. While the ETM, like LDA, is limited to the analysis of static corpora, it can also be extended to incorporate dynamic assumptions. This is the aim of the dynamic embedded topic model (DETM) that we describe in the following section.

3   Methodology

In this section, we first describe our baseline, the dynamic embedded topic model (Sect. 3.1), and then we present the proposed approach (Sect. 3.2).

3.1   The Dynamic Embedded Topic Model

The dynamic embedded topic model (DETM) joins the benefits of DTM and ETM, allowing the model to capture the topics' evolution over time while leveraging the representational power of word embeddings. The dynamic assumption over the topic proportions is the same as for the DTM:

$$\alpha^t \sim \mathcal{N}(\alpha^{t-1}, \delta^2 I) \tag{2}$$

but a dynamic prior is now assumed over the topic embeddings:

$$\eta^t \sim \mathcal{N}(\eta^{t-1}, \gamma^2 I) \tag{3}$$

The rest of the generative model for slice t is:

- For the d-th document, draw its topic proportions (logit scale):
    - $\theta_d \sim \mathcal{N}(\alpha^t, a^2 I)$.
- For each word in the d-th document:
    - Draw its topic: $z_{d,n} \sim \mathrm{Cat}(\sigma(\theta_d))$
    - Draw the word from the topic's word distribution:
    - $w_{d,n} \sim \mathrm{Cat}(\sigma(\rho^\top \eta_{z_{d,n}}^t))$.

The training of DETM involves maximizing the posterior distribution over the model's latent variables, $p(\theta, \eta, \alpha | D)$. However, maximizing the exact posterior is intractable. Therefore, the common approach is to approximate it with variational inference [6] using a factorized distribution, $q_v(\theta, \eta, \alpha | D)$. Its parameters, noted collectively as v, are optimized by minimizing the Kullback-Leibler (KL) divergence between the approximation and the posterior, which is equivalent to maximizing the following expectation lower bound (ELBO):

$$\mathcal{L}(v) = \mathbb{E}[\log p(\theta, \eta, \alpha, D) - \log q_v(\theta, \eta, \alpha | D)] \tag{4}$$

The implementation of q_v relies on feed-forward neural networks to predict the variational parameters, and on LSTMs to capture the temporal dependencies; we refer the reader to [10] for details.

3.2   The Proposed Approach: DETM-tau

The fundamental evaluation measure for a topic model is the *topic coherence* [14]. This measure looks at the "top" words in the word distribution of each topic, and counts how often they co-occur within each individual document. The assumption is that the higher the co-occurrence, the more "coherent" is the extracted topic model.

However, topic models cannot be trained to optimize the topic coherence. In the first place, the coherence is a counting measure that depends on the outcome of a ranking operation (a top-K argmax), and it is therefore not differentiable in the model's parameters. In the second place, it is evaluated globally over the entire document set. As a consequence, alternative approaches based on reinforcement learning [23] would prove excruciatingly slow, and would not be able to single out and reward the contribution of the individual documents (the so-called "credit assignment" problem [17]) (Table 1).

For this reason, in this work we attempt to improve the topic coherence by utilizing a softmax *with temperature* [20] in the word distributions. The inclusion of a temperature parameter can make the word distributions "sharper" (i.e. the probability mass more concentrated in the top words, for temperatures <1) or smoother/more uniform (for temperatures >1). We expect this to have an impact on the final word ranking, as high temperatures will make mixing more pronounced during training, while low temperatures may "freeze" the ranking

Table 1. Key sizes of the datasets used for the experiments.

Dataset	Training set	Validation set	Test set	Timestamps	Vocabulary
CORD-19TM	15,300	900	1,800	18	70,601
UNGDC	1,96,290	11,563	23,097	46	12,466
ACL	8,936	527	1,051	31	35,108

to an extent. With the addition of the temperature parameter, τ, the word distributions take the form:

$$w \sim \mathrm{Cat}(\sigma(\rho^{\top}\eta_z/\tau)) \tag{5}$$

While parameter τ can be optimized with the training objective like all the other parameters, we prefer using a simple validation approach over a small, plausible range of values to select its optimal value.

4   Experiments and Results

4.1   Experimental Set-Up

For the experiments, we have used three popular document datasets: the COVID-19 Open Research Dataset (CORD-19) [24], the United Nation General Debate Corpus (UNGDC) [3] and the ACL Title and Abstract Dataset (ACL) [5]. CORD-19 is a resource about COVID-19 and related coronaviruses such as SARS and MERS, containing over 500,000 scholarly articles, of which 200,000 with full text. For our experiments, we have created a subset organized in monthly time slices between March 2020 and August 2021, limiting each slice to the first 1,000 documents in appearance order to limit the computational complexity. We refer to our subset as CORD-19TM, and we release it publicly for reproducibility of our experiments. UNGDC covers the corpus of texts of the UN General Debate statements from 1970 to 2015 annotated by country, session and year. For this dataset, we have considered yearly slices. The ACL dataset [5] includes 10,874 title and abstract pairs from the ACL Anthology Network which is a repository of computational linguistics and natural language processing articles. For this dataset, too, we have considered yearly slices, with the years spanning from 1973 to 2006 (NB: three years are missing). As in [10], the training, validation and test sets have been created by splitting the datasets into 85%, 5% and 10% splits, respectively. All the documents were preprocessed with tokenization, stemming and lemmatization, eliminating stop words and words with document frequency greater than 70% and less than 10%, as in [10].

As models, we have compared the proposed DETM-tau with: the original DETM, DTM, and LDA applied separately to each individual time slice. As performance metrics, we have used the *perplexity* and the *topic coherence* which are the de-facto standards for this task. The perplexity is a measure derived from the probability assigned by the model to a document set, and should be as low as possible. It is typically measured over the test set to assess the model's generalization. The topic coherence is a measure of the co-occurrence of the "top" K words of each topic within single documents, and should be as high as possible. It is typically measured over the training set to assess the explanatory quality of the extracted topics. Several measures for the topic coherence have been proposed, and we use the NPMI coherence [14] with $K = 10$, as in [10]. As number of topics, we have chosen 20 and 40 which are commonly-used values in the literature. For the selection of the temperature parameter, τ, we have used range [0.25–2.25] in 0.5 steps. All other hyperparameters have been left as in the corresponding original models.

4.2   Results

Tables 2 and 3 show the results over the CORD-19TM dataset with 20 and 40 topics, respectively. In terms or perplexity, the proposed DETM-tau has neatly outperformed the original DETM for both 20 and 40 topics (NB: the perplexity is not available for the LDA and DTM models). In terms of topic coherence, DETM-tau has, again, achieved the highest values. The second-best results have been achieved in both cases by DTM, while DETM and LDA have reported much lower scores. In particular, the very poor performance of LDA shows that applying a standard topic model separately on each time slice is an unsatisfactory approach, and musters further support for the use of dynamic topic models for timestamped document analysis.

Tables 4 and 5 show the results over the UNGDC and ACL datasets, respectively. For these datasets, we have not carried out experiments with DTM as it proved impractically time-consuming, and we omitted LDA outright because of its non-competitive performance. On both these datasets, too, DETM-tau has been able to achieve both lower perplexity and higher coherence than the original DETM. We believe that these results provide clear evidence of the importance of controlling the sharpness-smoothness trade-off of the word distributions.

To explore the sensitivity of the results to the temperature parameter, τ, Fig. 1 plots the values of the perplexity and the topic coherence of DETM-tau (CORD-19TM, 20 topics) for various values of τ, using DETM as the reference. It is clear that setting an appropriate value is important for the model's performance. However, the plots show that the proposed model has been able to outperform DETM for an ample range of values. In addition, Fig. 2 plots the values of the perplexity and the topic coherence at successive training epochs. The plots show that both metrics improve for both models as the training progresses. Given that the topic coherence is not an explicit training objective, its increase along the epochs is remarkable and gives evidence to the effective design of both models.

Eventually, we present a concise qualitative analysis of the extracted topics through Table 6 and Fig. 3. Table 6 shows a few examples of the topics extracted by DETM-tau from the CORD-19TM dataset (20 topics) at time slices 0, 10 and 17. Each topic is represented by its ten most frequent words. Overall, all the examples seem to enjoy good coherence and descriptive power. For instance, the

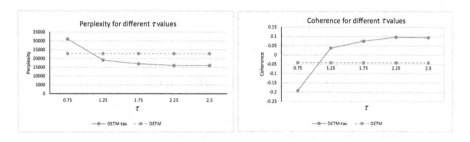

Fig. 1. Perplexity and topic coherence for DETM-tau for various values of the temperature parameter, τ (CORD-19TM, 20 topics). The value for DETM is used for comparison.

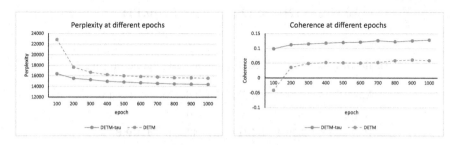

Fig. 2. Perplexity and topic coherence for DETM and DETM-tau at successive training epochs (CORD-19TM, 20 topics).

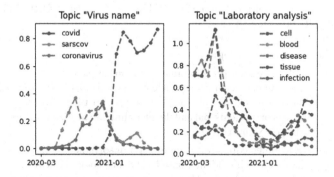

Fig. 3. Evolution of the probability of a few, selected words within their topics for the DETM-tau model with the CORD-19TM dataset, 20 topics.

Table 2. Results on the CORD-19TM dataset with 20 topics

Model	LDA	DTM	DETM	DETM-tau
Perplexity	–	–	15548.8	**14379.2**
Coherence	−0.049	0.114	0.059	**0.129**

Table 3. Results on the CORD-19TM dataset with 40 topics

Model	LDA	DTM	DETM	DETM-tau
Perplexity	–	–	14966.3	**13129.7**
Coherence	−0.047	0.081	−0.043	**0.093**

Table 4. Results on the UNGDC dataset with 20 and 40 topics

Model	DETM	DETM-tau	DETM	DETM-tau
# topics	20		40	
Perplexity	3032.8	**3023.5**	2798.9	**2782.0**
Coherence	0.121	**0.129**	0.048	**0.124**

Table 5. Results on the ACL dataset with 20 and 40 topics

Model	DETM	DETM-tau	DETM	DETM-tau
# topics	20		40	
Perplexity	5536.4	**5421.1**	4360.0	**4169.6**
Coherence	0.150	**0.179**	0.153	**0.174**

Table 6. Examples of topics extracted by DETM-tau from the CORD-19TM dataset (20 topics) at different time slices.

Time slice	Examples of topics
0	Zikv cytokine proinflammatory resuscitation ferritin antitumor exosomes thoracic evidence based patienten cells infection cell virus blood disease protein tissue infected receptor patients patient health clinical care hospital months disease years therapy
10	Exosomes copd frailty mgml tavi absorbance biofilm sigmaaldrich evidence based virulence social education research health people services industry culture educational providers macrophages antibacterial antioxidant kshv mmp lmics propolis sdgs inactivation hydrogel patients studies health care patient clinical treatment disease population risk
17	Nanoparticles proinflammatory bioactive antifungal inhospital coagulation angiogenesis inflammasome cells cell blood disease tissue cancer infection protein proteins metabolism patients health patient social education hospital clinical people care population

first topic at time slice 0 could be titled "immune response analysis" or something akin; the last topic at time slice 17 could be titled "population health"; and so forth. Therefore, the automated categorization of the articles into such topics seem to provide a useful, and completely unsupervised, analysis. In turn, Fig. 3 shows the temporal evolution of the frequency of a few, manually selected words within their respective topics. The left-most topic, which we have labelled as "virus name", shows that referring to COVID-19 by the names "coronavirus" and "sarscov" was popular during 2020; conversely, as of January 2021, the name "covid" has become dominant. The right-most topic shows that words such as "blood", "infection" and "tissue" have decreased their in-topic frequency over time, possibly in correspondence with an increased understanding of the disease. These are just examples of the insights that can be obtained by dynamic topic models.

5   Conclusion

This paper has presented a temperature-modified dynamic embedded topic model for topic modelling of timestamped document collections. The proposed model uses a softmax with temperature over the word distributions to control their sharpness/smoothness trade-off and attempt to achieve a more effective parameterization of the overall topic model. Experiments carried out over three timestamped datasets (a subset of the CORD-19 dataset referred to as CORD-19TM, the United Nation General Debate Corpus (UNGDC) and the ACL Title and Abstract Dataset (ACL)) have showed that the proposed model, suitably nicknamed DETM-tau, has been able to outperform the original DETM model by significant margins in terms of both perplexity and topic coherence. In addition, DETM-tau has performed remarkably above the other compared models. A qualitative analysis of the results has showed that the proposed model has generally led to interpretable topics, and can offer insights into the evolution of the topics over time.

References

1. Alvarez-Melis, D., Saveski, M.: Topic modeling in Twitter: aggregating tweets by conversations. In: The 10th International Conference on Web and Social Media, pp. 519–522 (2016)
2. Arnold, C., El-Saden, S., Bui, A., Taira, R.: Clinical case-based retrieval using latent topic analysis.In: AMIA Annual Symposium Proceedings, vol. 2010, pp. 26–30 (2010)
3. Mikhaylov, S.J., Baturo, A., Dasandi, N.: Understanding state preferences with text as data. In: Introducing the UN General Debate Corpus. Research & Politics (2017)
4. Bengio, Y., Ducharme, R., Vincent, P., Janvin, C.: A neural probabilistic language model. J. Mach. Learn. Res. **3**, 1137–1155 (2003)
5. Bird, S., et al.: The ACL anthology reference corpus: a reference dataset for bibliographic research in computational linguistics. In: International Conference on Language Resources and Evaluation, pp. 1755–1759 (2008)

6. Blei, D.M., Kucukelbir, A., McAuliffe, J.D.: Variational inference: a review for statisticians, pp. 859–877 (2017)
7. Blei, D.M., Ng, A.Y., Jordan, M.I.: Latent Dirichlet allocation. J. Mach. Learn. Res. **3**, 993–1022 (2003)
8. Cecchini, M., Aytug, H., Koehler, G.J., Pathak, P.: Making words work: using financial text as a predictor of financial events. Decis. Support Syst. **50**(1), 164–175 (2010)
9. Devyatkin, D., Nechaeva, E., Suvorov, R., Tikhomirov, I.: Mapping the research landscape of agricultural sciences. Foresight STI Govern. **12**(1), 57–76 (2018)
10. Dieng, A.B., Ruiz, F.J.R., Blei, D.M.: The dynamic embedded topic model (2019)
11. Dieng, A.B., Ruiz, F.J.R., Blei, D.M.: Topic modeling in embedding spaces. Trans. Assoc. Comput. Linguist. **8**, 439–453 (2020)
12. Kim, H., Drake, B., Endert, A., Park, H.: ArchiText: interactive hierarchical topic modeling. IEEE Trans. Vis. Comput. Graphics **27**(9), 3644–3655 (2021)
13. Lafferty, J.D., Blei, D.M.: The dynamic topic model. In: The 23rd International Conference on Machine Learning, pp. 113–120 (2006)
14. Lau, J.H., Newman, D., Baldwin, T.: Machine reading tea leaves: automatically evaluating topic coherence and topic model quality. In: The 14th Conference of the European Chapter of the Association for Computational Linguistics (EACL 2014), pp. 530–539 (2014)
15. Le, Q.V., Mikolov, T.: Distributed representations of sentences and documents. In: The 31th International Conference on Machine Learning, vol. 32, pp. 1188–1196 (2014)
16. Liu, T., Zhang, N.L., Chen, P.: Hierarchical latent tree analysis for topic detection. CoRR, vol. 8725, pp. 256–272 (2014)
17. Minsky, M.: Steps toward artificial intelligence. Proc. IRE **49**(1), 8–30 (1961)
18. Nguyen, T.H., Shirai, K.: Topic modeling based sentiment analysis on social media for stock market prediction. In: The 53rd Annual Meeting of the Association for Computational Linguistics and the 7th International Joint Conference on Natural Language Processing of the Asian Federation of Natural Language Processing (ACL-IJCNL 2015), pp. 1354–1364 (2015)
19. Peng, M., et al.: Neural sparse topical coding. In: The 56th Annual Meeting of the Association for Computational Linguistics (ACL 2018), pp. 2332–2340 (2018)
20. Platt, J.C.: Probabilistic outputs for support vector machines and comparisons to regularized likelihood methods. In: Advances in Large Margin Classifiers, pp. 61–74. MIT Press (1999)
21. Rodrigues, F., Lourenco, M., Ribeiro, B., Pereira, F.C.: Learning supervised topic models for classification and regression from crowds. IEEE Trans. Pattern Anal. Mach. Intell. **39**(12), 2409–2422 (2017)
22. Sarioglu, E., Choi, H.-A., Yadav, K.: Clinical report classification using natural language processing and topic modeling. In: The 11th International Conference on Machine Learning and Applications, vol. 2, pp. 204–209 (2012)
23. Sutton, R.S., Barto, A.G.: Reinforcement Learning: An Introduction, 2nd edn. (2018)
24. Wang, L.L., et al.: CORD-19: the COVID-19 open research dataset. In: 1st Workshop on NLP for COVID-19 at ACL 2020, vol. 1, pp. 1–12 (2020)
25. Guixian, X., Meng, Y., Chen, Z., Qiu, X., Wang, C., Yao, H.: Research on topic detection and tracking for online news texts. IEEE Access **7**, 58407–58418 (2019)
26. Zhang, A., Zhu, J., Zhang, B.: Sparse online topic models. In: The 22nd International World Wide Web Conference (WWW 2013), pp. 1489–1500 (2013)

27. Zhang, R., Pakhomov, S., Gladding, S., Aylward, M., Borman-Shoap, E., Melton, G.: Automated assessment of medical training evaluation text. In: AMIA Annual Symposium Proceedings, vol. 2012, pp. 1459–68 (2012)
28. Zhu, J., Xing, E.P.: Sparse topical coding. In: The 27th Conference on Uncertainty in Artificial Intelligence (UAI 2011), pp. 831–838 (2011)

Measuring Difficulty of Learning Using Ensemble Methods

Bowen Chen[✉], Yun Sing Koh, and Ben Halstead

School of Computer Science, The University of Auckland, Auckland, New Zealand
bche264@aucklanduni.ac.nz, {y.koh,ben.halstead}@auckland.ac.nz

Abstract. Measuring the difficulty of each instance is a crucial meta-knowledge extraction problem. Most studies on data complexity have focused on extracting the characteristics at a dataset level instead of the instance level while also requiring the complete label knowledge of the dataset, which can often be expensive to obtain. At the instance level, the most commonly used metrics to determine difficult to classify instances are dependant on the learning algorithm used (*i.e.*, uncertainty), and are measurements of the entire system instead of only the dataset. Additionally, these metrics only provide information of misclassification in regard to the learning algorithm and not in respect of the composition of the instances within the dataset. We introduce and propose several novel instance difficulty measures in a semi-supervised boosted ensemble setting to identify difficult to classify instances based on their learning difficulty in relation to other instances within the dataset. The proposed difficulty measures measure both the fluctuations in labeling during the construction process of the ensemble and the amount of resources required for the correct label. This provides the degree of difficulty and gives further insight into the origin of classification difficulty at the instance level reflected by the scores of different difficulty measures.

Keywords: Complexity measures · Boosting · Instance difficulty

1 Introduction

All instances in a dataset have different degrees of classification difficulty. The classification difficulty of an instance is largely dependent on the learning algorithm used for classification and the composition of the instance within the dataset. There are two main approaches to measuring classification difficulty: model-centric approaches based on the uncertainty of the classifier [12,16,20], data-centric approaches in the form of data complexity measures that reflect the characteristics of the data [9,11,14]. These two approaches all have their advantages and limitations.

While being able to function without the complete label knowledge of the dataset, uncertainty-based difficulty measures are dependent on the learner;

Supplementary Information The online version contains supplementary material available at https://doi.org/10.1007/978-981-19-8746-5_3.

therefore the difficulty score outputs are measurements of the entire system instead of only the datasets. Consider an example of a linear classifier, *e.g.* Linear Support Vector Machine (Linear SVM), is used to set the decision boundary. Instances x_1, x_2, x_3, and x_4 are instances of the same class. Instances x_1 and x_4 are considered outliers; however, in this scenario, instance x_1 would almost always be correctly classified, and instance x_4 would almost always be incorrectly classified. Instances near the decision boundary, such as instances x_2 and x_3, would be more difficult to classify correctly than inlier instances further away from the decision boundary. If a different learner is used, such as a non-linear classifier may produce a more complex decision boundary, then the classification difficulty of instances around the decision boundary will change. For example, instance x_2 may become harder to classify, and instance x_3 may become easier to classify given a more complex decision boundary due to the decision bias from the learner. A solution for this is to have a set of diverse learning algorithms to generalize beyond a single learning algorithm [20]. However, this can be computationally expensive due to the number of different learning algorithms, limiting the application cases.

On the other hand, difficulty scores based on data complexity are independent of the learner, but they have the disadvantages of requiring the complete label knowledge of the dataset and are limited at an instance level as most of the measures only characterize a dataset's overall complexity but cannot give indications of instance complexity. At the instance level, the majority of previous work has focused on specific cases of difficult to classify instances such as outliers (x_4) [1,21], border points (x_2, x_3) [3,13], or minority classes [2,6]. As these methods focus on individual cases, it is possible that not all inconsistently labeled instances can be identified.

We hypothesize that in some applications, a more accurate method of instance-level classification difficulty calculation is to have a combination of both approaches, where the score is independent of the learner and functional with limited data. Motivated by the method of creating a generalized learning algorithm by having a set of learners [20] and the fact that boosting performs poorly with respect to noisy instances [5], we use an ensemble boosting approach to estimate the classification difficulty at an instance level, precisely, AdaBoost SAMME (Stagewise Additive Modeling using a Multi-class Exponential loss function) [10], which incrementally builds an ensemble of weak classifiers of decision stumps. In the following sections, we describe the setting of our proposed method and then a set of difficulty measures based on this approach. Figure 1 shows an overview of our proposed framework.

2   Related Work

A number of methods have been proposed to identify instances that are difficult to classify. The dataset complexity measures from Ho and Basu [11] has been widely used to analyze the classification difficulty problem. Ho and Basu noted that the classification difficulty mainly originates from a combination of three

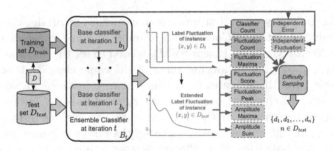

Fig. 1. Overview of the proposed framework

sources: 1) Class ambiguity which is when for a given dataset the classes cannot be distinguished regardless of learning algorithm used; 2) The dimensionality and sparsity of the dataset; and 3) The complexity of the decision boundary. In addition, they present a number of measures that focus on the complexity of the classification boundary such as the Fisher's Discriminant Ratio (F1). This idea of class ambiguity is also shared in the recent work of Hullermeier and Waegeman [12], at which they argue the distinction between two sources of uncertainty: aleatoric and epistemic. Similar to class ambiguity, aleatoric uncertainty is the irreducible uncertainty where even with the precise knowledge of the optimal hypothesis, the prediction is still uncertain, and epistemic uncertainty is reducible uncertainty caused by a lack of data. Extending from Ho and Basu's work, several other complexity measure that capture similar aspects of complexity has been proposed [9,14,20]. However these methods all assume the availability of labelled data and are designed for supervised classification problems, limiting their applicability in areas with limited labeled data such as active learning.

3   Instance Difficulty

Given a dataset $D = \{(x_1, y_1), \ldots, (x_n, y_n)\}$ of size n with n_c unique class labels, the classification difficulty d_i of an instance x_i is defined as the amount of resources required for a learner θ to learn the underlying concept and output the ground truth label y_i. We measure both time and memory computation resources. For our proposed method, we train one weak base leaner as the unit for measuring the classification difficulty of an instance. Specifically, we deliberately overfit to difficult to classify instances and measure the number of weak learners required for an instance to be correctly classified. This process of overfitting by including more weak learners is fundamentally an increase in time and space resources.

AdaBoost [7] is chosen over other boosting methods, whereas, other boosting methods [4,8] do not penalize misclassified instances but instead use a loss function. Overall Adaboost model consists of t_{total} weak learners $B_{t_{total}}(x) = \sum_{t=1}^{t_{total}} b_t(x)$ where b_t is the t weak learner that takes instance x as input and returns a score indicating the predicted label of the instance. The boosting process consists of a number of iterations where a weak classifier is boosted and added to the ensemble at each iteration using information gained from previous iterations.

Theoretically, as the number of weak classifiers increases, the amount of performance gained from each additional weak learner decreases. Following the definition of classification difficulty, difficult instances would need more resources to be correctly labeled. For each instance, we monitor the changes in the label, correctly or incorrectly labeled denoted by 0 and 1 respectively, as new weak classifiers are added. Given a Adaboost model $B_{t_{total}}$ consisting of $b_1, \ldots, b_{t_{total}}$ weak learners. The label fluctuation sequence of an instance x_i is $fluc_{t_{total}}(x_i)$ denoted by $fluc_{t_{total}}(x_i) = \{f_t\}_{t=1}^{t_{total}}(x_i) = \{f_1(x_i), \ldots, f_{t_{total}}(x_i)\}$ where $f_t(x_i)$ is 1 if x_i is incorrectly labeled by the ensemble model at iteration t and 0 otherwise. Here $f_t(x_i) = I(B_t(x_i))$ where $I(\cdot)$ is an indicator function:

$$I(f(x_i)) = \begin{cases} 0, & \text{if } f(x_i) == y_i \\ 1, & \text{otherwise.} \end{cases} \tag{1}$$

Intuitively more difficult to classify instances will require more weak base classifiers to be correctly classified than easy ones. In the next section, we present nine difficulty measures that measure the following: the computation resources required for a correct and consistent label, the consistency of the label, and the state of the ensemble when the label is misclassified. All proposed difficulty measures are aimed to extract meta-information that provides indications of classification difficulty from the fluctuations in label; hence these measures are not limited to only labeled data.

3.1   Difficulty Measures

This section introduces three difficulty measures that extract meta-information and provide indications of classification difficulty based on the fluctuations in the label. We also distinguish between two types of classification difficulty, namely, label difficulty and correctness difficulty.

Base Classifier Count C. This measure approximates the minimum computation required to correctly label an instance, which is similar to estimating an instance's description length. The base classifier count of an instance is the number of weak classifiers necessary to build an ensemble that can predict the correct label, denoted by the following:

$$C(x) = \frac{\text{argmin}_t B_t(x) == y, B_t(x) \in B_{t_{total}}(x)}{t_{total}} \tag{2}$$

where y is the true label of x, and t_{total} is the total number of weak classifiers.

Fluctuation Count: F_{cnt}. In addition to the minimum computation resource required for the correct label, the consistency of an instance's label also provide an indication of classification difficulty. This metric measures the uncertainty of

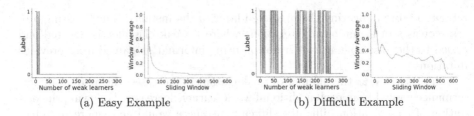

(a) Easy Example   (b) Difficult Example

Fig. 2. The label fluctuation sequence of an easy to classify instance (a) and a difficult to classify instance (b)

an instance's predicted label based on the fluctuations in the label as more weak learners are added to the overall ensemble model and is given by:

$$F_{cnt}(x) = \frac{\sum_{t=1}^{t_{total}} I^f(f_t(x), f_{t-1}(x))}{t_{total}} \tag{3}$$

where $f_t(x)$ is the tth label fluctuation sequence of an instance and $I^f(\cdot)$ is a indicator function that return 1 if there is a change in label, it is denoted as:

$$I^f(f(x_i), g(x_i)) = \begin{cases} 0, & \text{if } f(x_i) == g(x_i) \\ 1, & \text{otherwise} \end{cases} \tag{4}$$

If an instance experiences many label fluctuations, it indicates the instance is close to instances of other classes, and the label of the instance is influenced by the additional hypothesis input from a newly added weak learner. Figure 2 shows the fluctuation graph of an example of an easy to classify instance with 6 fluctuations in the label and an example of a more difficult to classify instance with 43 fluctuations.

Fluctuation Score F. The Fluctuation Score is a measurement of label consistency. From the fluctuations in the label, we apply a sliding window to produce a fluctuation graph such that small label fluctuations contribute less to the overall fluctuations in label. Given a fluctuation sequence $fluc_{t_{total}}$, the extended fluctuation sequence is a series containing averages of different subsets of the fluctuation sequence. Each subset is defined by a sliding window that moves across the fluctuation sequence. The size of the sliding window increases at the beginning of the sequence up to a predefined window size and decreases at the endpoints such that the average is taken over only the elements that fill the window. The extended fluctuation sequence with a sliding window of size l is defined as $S_l(fluc_{t_{total}}(x)) = \{s_{1,1}, s_{1,2}, \ldots, s_{1,l}, s_{2,l}, \ldots, s_{t_{total}-1,2}, s_{t_{total},1}\}$ where $s_{t,l}$ is the window average of the tth fluctuation in the label fluctuation sequence, with a window size of l, and it is given by:

$$s_{t,l} = \frac{1}{l} \sum_{j=t}^{t+l-1} f_j, f_j \in fluc_{t_{total}}(x). \tag{5}$$

Figure 2 shows extended fluctuation graph, where the sliding window size is the same as the size of the fluctuation sequence ($t_{total} = 300$), and the window size increases from 1 to 300, then back down to 1, resulting in an extended fluctuation sequence that contains 600 window averages. The fluctuation score of an instance is a ratio between the area under curve and the area above curve of the fluctuation graph given by:

$$F(x) = 1 - \text{abs} \left(\frac{2 \times \sum S_l(x)}{|S_l(x)|} - 1 \right). \tag{6}$$

The F value has a range between $[0, 1]$. A value closer to 1 indicates that the instance is correctly classified and misclassified for a similar number of iterations and the label of the instance is inconsistent. If the instance is labeled correctly classified and incorrectly classified for the same duration, the fluctuation score will be zero.

Fluctuation Maxima: F_{max} This measure provides an indication of the state of the ensemble model when a given instance experiences the most misclassification. The institution is that instances that are frequently incorrectly labeled when the ensemble model is over-fitting to difficult instances, consisting of more weak learners, will be more difficult to classify compared to instances that are misclassified when the ensemble model is weaker with a fewer number of weak learners. Fluctuation Maxima is the size of the ensemble when the instances remain misclassified for the longest number of iterations. The pseudo code can be found in the supplementary materials. A higher F_{max} value indicates that the instance experiences the most misclassification when the ensemble is stronger, therefore, more difficult to classify. Vice versa, a lower F_{max} value is indicative of an easier to classify instance. Given that the F_{max} value is normalized in respect to the number of weak learners, F_{max} has a range of $[0, 1]$

Fluctuation Peak: F_{peak} The Fluctuation Peak metric, similar to F_{max}, also measures the ensemble state when the instance experience the most misclassification. However, instead of looking only at the position of most misclassification given the fluctuations sequence, fluctuation peak is the magnitude of the vector consisting of all peak positions given by the extended fluctuation sequence $S_l(fluc_{t_{total}}(x))$, and it is expressed as:

$$F_{peak}(x) = \|Peaks(S_l(fluc_{t_{total}}(x)))\| \tag{7}$$

where the function $Peaks(\cdot)$ returns the indices of all peak positions, each peak in the extended fluctuation sequence indicates a change in the label that remained for a significant number of iterations. Therefore, instances with a high number of peaks when the ensemble consists of a high number of weak learners will result in a high F_{peak} value. These instances are considered more difficult to classify.

Maximum Amplitude: A_{max} Maximum Amplitude measures the size of the most significant fluctuation within the graph in the frequency amplitude domain. Fast Fourier Transform (FFT) [17] is applied to the extended fluctuation sequence of each instance, from which we plot a frequency spectrum and take the max amplitude. The Maximum Amplitude can be denoted by:

$$A_{max}(x) = \max(FFT(S_l(fluc_{t_{total}}(x))))\qquad(8)$$

where the function $FFT(.)$ returns the Fourier transformation of its argument. By applying the Fourier transformation to the extended fluctuation sequence, we can decompose the sequence which initially is in the time domain into the frequency domain. We can then obtain the amplitude of the label fluctuations from the transformed output by taking their respective absolute values. In this case, difficult to classify instances are instances that have high amplitudes given any frequencies. It indicates the instance has been misclassified for an extended number of iterations, and the instance is more difficult to classify.

Amplitude Sum: A_{sum} Amplitude Sum measures the size of all fluctuations instead of only the most significant fluctuation, and it is given by:

$$A_{sum}(x) = \sum FFT(S_l(fluc_{t_{total}}(x))))\qquad(9)$$

Similarly, FFT is applied to the extended fluctuation graph of each instance, from which we take the sum of all amplitudes regardless of frequency. In this case, an instance is considered to be the most difficult to classify if it has multiple extended label fluctuations, indicated by a high Amp_{sum} value. A_{sum} also has a range between 0 and $+\inf$.

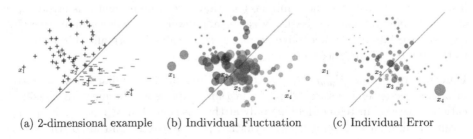

(a) 2-dimensional example   (b) Individual Fluctuation   (c) Individual Error

Fig. 3. Individual fluctuation and individual error score of the example from Figure (a). The score of each instance is represented by the size of the circles.

Individual Fluctuation Score IF. Instead of the cumulative prediction by the ensemble model $B_{t_{total}}$, the prediction of the weak learner b_t is used to construct the individual fluctuation sequence of an instance denoted by $fluc_{t_{total}}(x_i)$, and it is given by: $fluc_{t_{total}}(x_i) = \{\hat{f}_t\}_{t=1}^{t_{total}}(x_i) = \{\hat{f}_1(x_i), \ldots, \hat{f}_{t_{total}}(x_i)\}$, where $\hat{f}_t(x_i)$ is 1 if x_i is incorrectly labeled by the weak learner at iteration t and 0 otherwise. It is denoted by: $\hat{f}_t(x_i) = I(b_t(x_i))$

Differently to F_{cnt}, the Individual Fluctuation Score measures the number of times an instance's label changes for each individual weighted weak classifier that makes up the ensemble. Note that IF is given in regards to t_{total} and ranges between 0 and 1. The IF value is computed as follows:

$$IF(x) = \frac{\sum_{t=1}^{t_{total}} I^f(\hat{f}_t(x), \hat{f}_{t-1}(x))}{t_{total}} \tag{10}$$

where $I^f(\cdot)$ is the same as in Eq. 4, and return 1 if there is a change in label. Figure 3b shows the IF value of instances from Fig. 3a denoted by the size of the circles. Instances that are closer to the decision boundary have a higher F_{cnt} value, as these instances experience more label fluctuations. When a base classifier tries to label an weighted instance that was previously incorrectly labeled correctly, the base classifier also affects the labels of nearby samples with similar features, which causes them to have an increase in fluctuation count.

Individual Error Score IE. Individual Error Score measure the number of times an instance is incorrectly labeled by each individual weak classifier of the ensemble defined as:

$$IE(x) = \frac{\sum_{t=1}^{t_{total}} I(b_t(x))}{t_{total}}. \tag{11}$$

Each weak classifier of the ensemble has different sample weights $w_{i,t}$ determined by the previous weak classifier. Intuitively more difficult instances with a higher IE value will be misclassified more often compared to easier instances.

Label Difficulty Vs Correctness Difficulty. Similar to epistemic uncertainty and aleatoric uncertainty [12], a distinction can be made between label difficulty and correctness difficulty. Label difficulty is the difficulty of correctly labeling an instance, and similar to epistemic uncertainty, this difficulty can be reduced if provided with enough relevant data. On the other hand, correctness difficulty is the difficulty in determining if an instance is correctly labeled or incorrectly labeled. Both an instance that is incorrectly classified for all iterations and another that is correctly classified for all iterations are considered to have low correctness difficulty as we can be confident that the label is either correct or incorrect. This is similar to aleatoric uncertainty, where the difficulty cannot be reduced with additional data. In our case, the label fluctuations-based difficulty measures F is a measurement of correctness difficulty and cannot distinguish between an instance that is incorrectly classified for all the iterations and another that is correctly classified for the entire duration. In comparison, the classifier counts C as a metric of label difficulty and can distinguish between two such instances. The individual error scores IE while being a measurement label difficulty, it is affected by nearby high fluctuation instances and indicates correctness difficulty.

Pseudo True Label Assumptions. Given that the difficulty measures are based on label fluctuations, in the absence of labeled data we can assume 1) the most probable label or 2) the most consistent label as the pseudo true label.

The most probable label is the label with the highest prediction probability given by $\hat{y} = \{y | Pr(y) = max(Pr(y))\}$. The most consistent label is the label that remains the same for the longest number of iterations during the boosting process. We monitor the fluctuation between correct (0) and incorrect labels (1). We are interested in the fluctuation between all possible labels. We denote the prediction sequence as $P_{t_{total}}(x_i) = \{p_t\}_{t=1}^{t_{total}}(x_i) = \{B_t(x_i), \ldots, B_{t_{total}}(x_i)\}$ where p_t is the prediction output of the ensemble model with t weak learners. The value range of the prediction sequence is dependent on the number of unique class labels, for example, the most consistent label of an instance with a prediction sequence of $\{0, 1, 1, 1, 2, 0, 0\}$ is labelled 1 as the label is maintained for three continuous iterations.

Complexity Analysis Discussion. Our method consists of an ensemble boosting algorithm and some difficulty measure calculations. Given a dataset D with training set D_X, the time complexity of training a decision stump base classifier b_t is $\mathcal{O}(|D_X| \times f)$ where f is the number of features. The cost of training an ensemble with t base classifiers is therefore $\mathcal{O}(t \times |D_X| \times f)$ and the test time complexity is $\mathcal{O}(t \times f)$. In addition to the ensemble testing cost, given that all difficulty measures are based on the label fluctuations, each difficulty measure can be computed in linear time in respect to t. Additionally, the training cost of the ensemble can be effectively reduced by scaling down the number of base learners included in the ensemble.

4   Experiments

We perform several experiments to answer the following questions: (RQ 1) How does the proposed difficulty correlate with each other? and are there possible overlaps in what they measure? (RQ 2) How does the proposed difficulty correlate with existing uncertainty measures? (RQ 3) How does the proposed difficulty correlate with existing complexity measures?

Datasets. Experiments are conducted on both controlled synthetic and real-world datasets. Given that class skew magnifies any sources of classification difficulty present in the instances [18], we used random tree and random RBF data generators [15] to build a series of imbalanced datasets such that 75% of the dataset are instances belonging to the same class, and instances from other classes make up the remainder 25% of the dataset. A series of real-world datasets from the UCI repository are also used in our experiments. The result tables below only shows a portion of the results, the full result can be found in the supplementary materials and our code is available at https://anonymous. 4open.science/r/AUSDM22-1808/.

Evaluation Metrics. Spearman correlation [19] is used to compare the difficulty measures. A Spearman coefficient has a range from -1 to $+1$. A strong monotonic relationship is denoted by a value close to -1 or $+1$, and a value of 0 indicates that there is a monotonic association.

Baseline Methods. We use the misclassification frequency ($Freq$) and the classifier scores (Clf) from a set of ESLAs as the baseline comparison for uncertain-based classification difficulty measures [20]. Given that class overlap and class imbalance are the two main contributors to classification difficulty. We use the Intra/Inter ratio and the class imbalance ratio as the data complexity baselines. The Intra/Inter ratio for an instance (x_i, y_i) for a given dataset denoted by dNN is calculated as follows:

$$dNN(x_i, y_i) = \frac{d(x_i, NN(x_i) \in \{y_i\})}{d(x_i, NN(x_i) \notin \{y_i\})} \tag{12}$$

$d(x_i, NN(x_i) \in \{y_i\})$ is the distance between instance x_i to its nearest neighbour $NN(x_i)$ from the same class y_i and $d(x_i, NN(x_i) \notin \{y_i\})$ corresponds to the distance of x_i to the closest neighbor from another class $\notin \{y_i\}$.

The instance-level class imbalance ratio, CB, of an instance (x_i, y_i) is the ratio between the number of instances that have the same class to the number of instances in the majority class in dataset D, and it is given by:

$$CB(x_i, y_i) = \frac{|\{x_z | y_z == y_i, x_z \in D\}|}{Majority(D)} \tag{13}$$

where $Majority(D)$ is the number of instances in the majority class of the dataset D given by:

$$Majority(D) = \max_{y \in n_c}(|\{x_j | y_j == y, x_j \in D\}|). \tag{14}$$

Parameter Settings. The total number of base weak learners(t_{total}) for boosting is set to 300, and the sliding window size l is also set to 300 for extended fluctuation sequence calculations.

(RQ 1). Given that there are difficulty measures that measure the same difficulty property of an instance by design, we first calculate the difficulty measures of all instances from 26 real-world datasets and 10 generated synthetic datasets and examine the relationship between the hardness measures using five-by-five-fold cross-validation. This provides insight into the similarity between the difficulty measures and detects possible overlaps of what the difficulty measures are measuring. Figure 4 shows the pairwise comparison of all nine difficulty measures.

From the correlation matrix we can see that all difficulty measure based on the extended fluctuation sequence $S_l(x)$, (e.g. F, F_{cnt}, F_{peak}, A_{max} and A_{sum}) are highly correlated. This indicates these difficulty measures measure similar aspects of classification difficulty. On the other hand, difficulty measures C, IF, and IE have a weaker correlation with each other and fluctuation-based measures, which suggest that they are metrics of different sources of classification difficulty. Out of the nine difficulty measures, the Base Classifier Count measure and Fluctuation Maxima are strongly correlated, as they are both metrics

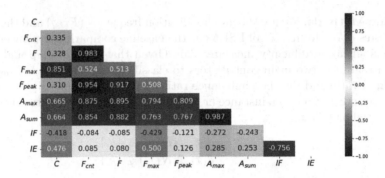

Fig. 4. Spearman correlation matrix of the difficulty measures

estimating the computational resource required for correct classifications. Interestingly, the Individual Fluctuation measure is negatively correlated with all other metrics, which suggests a lower IF value is indicative of a more difficult to classify instance as opposed to a higher IF value. The IF measure is calculated with respect to the number of weak learners included in the ensemble model, and for extremely difficult to classify instances, it is likely that these instances will remain incorrectly labeled for an extended number of iterations as opposed to fluctuating between incorrect and correct labels which results in a low IF value. In addition to the correlation between the proposed difficulty measure, we also examine the correlation with existing instance-level complexity measure in the following section to show that the proposed difficulty measures are measurements of classification difficulty.

(RQ 2). Table 1 shows the correlations of the nine difficulty measures with the uncertainty-based classification difficulty measures $Freq$ and Clf. The bold values represent the difficulty measures with the strongest correlation for each dataset. K denotes the kappa performance of the overall ensemble model for each dataset. The correlation results show that the proposed difficulty measures are correlated to uncertainty-based difficulty measures, with some datasets having a stronger correlation with some difficulty measures than others. For example, F_{max} has the strongest correlation for both $Freq$ and Clf with values 0.918 and 0.657 respectively for the RT1 Imb dataset, in comparison to all other difficulty measures. The difficulty measures for the Yeast dataset, on the other hand, are only weakly correlated to $Freq$ and Clf, even for the difficulty measure with the highest correlation IE, with correlation coefficients 0.321 and 0.250. This difference in correlation is expected as different datasets have different levels and sources of classification difficulty.

From the table, we can see that no one difficulty measure consistently has the strongest correlation with $Freq$ or Clf. This shows that difficulty measures with different levels of correlations are representative of different types of classification difficulty. For example, the classifier count score (C) has the highest correlation with both $Freq$ and Clf for the Codon dataset. In contrast, F_{peak} and IE have

Table 1. The correlations of proposed difficulty measures with uncertainty-based classification difficulty measures $Freq$ and Clf.

Dataset	Measure	C	F_{cnt}	F	F_{max}	F_{peak}	A_{max}	A_{sum}	IF	IE
Iris K: 0.916	Freq	0.413±0.054	0.394±0.051	0.519±0.032	0.453±0.064	0.190±0.057	**0.577±0.037**	0.552±0.054	−0.102±0.096	0.096±0.038
	Clf	0.558±0.142	0.580±0.131	0.689±0.041	0.602±0.139	0.339±0.104	**0.710±0.046**	0.683±0.047	0.223±0.166	0.350±0.101
Ecoli K: 0.719	Freq	**0.670±0.037**	0.423±0.053	0.490±0.037	0.641±0.044	0.375±0.048	0.561±0.041	0.532±0.038	−0.294±0.070	0.313±0.093
	Clf	0.476±0.032	0.568±0.047	0.606±0.035	**0.662±0.025**	0.537±0.036	0.634±0.040	0.572±0.044	−0.331±0.082	0.355±0.086
Car K: 0.678	Freq	0.576±0.014	0.544±0.025	0.549±0.021	**0.640±0.006**	0.529±0.024	0.596±0.012	0.568±0.011	0.050±0.036	0.500±0.009
	Clf	0.717±0.003	0.779±0.010	0.789±0.009	0.826±0.002	0.714±0.018	**0.827±0.002**	0.820±0.004	0.044±0.058	0.807±0.007
Spambase K: 0.875	Freq	0.190±0.017	0.103±0.013	0.100±0.016	0.237±0.017	0.132±0.012	0.168±0.020	0.148±0.023	0.111±0.013	**0.364±0.010**
	Clf	0.353±0.014	0.337±0.013	0.335±0.014	**0.404±0.007**	0.325±0.006	0.380±0.011	0.368±0.012	0.323±0.013	0.296±0.021
Yeast K: 0.316	Freq	0.190±0.017	−0.290±0.014	−0.282±0.019	0.117±0.040	−0.270±0.013	−0.272±0.018	−0.276±0.025	−0.271±0.041	**0.321±0.048**
	Clf	0.242±0.008	−0.040±0.024	−0.029±0.026	0.130±0.025	−0.035±0.027	−0.054±0.031	−0.102±0.020	−0.204±0.017	**0.250±0.025**
Abalone K: 0.117	Freq	0.391±0.025	−0.457±0.012	−0.454±0.014	0.002±0.038	−0.440±0.011	−0.437±0.022	−0.461±0.014	−0.257±0.023	0.310±0.018
	Clf	0.202±0.010	−0.162±0.020	−0.181±0.010	0.015±0.023	−0.155±0.014	−0.196±0.016	−0.225±0.014	−0.293±0.023	**0.320±0.025**
Nursery K: 0.635	Freq	**0.325±0.002**	0.196±0.019	0.119±0.022	0.299±0.020	0.189±0.021	0.118±0.020	0.088±0.014	−0.172±0.033	0.216±0.035
	Clf	**0.649±0.004**	0.531±0.028	0.334±0.032	0.419±0.031	0.528±0.033	0.306±0.032	0.231±0.024	−0.200±0.054	0.264±0.063
Adult K: 0.576	Freq	0.616±0.053	0.395±0.067	0.405±0.068	**0.620±0.052**	0.245±0.053	0.554±0.056	0.554±0.057	0.335±0.065	0.536±0.069
	Clf	0.391±0.021	0.351±0.028	0.337±0.025	0.423±0.017	0.295±0.023	0.393±0.016	0.388±0.017	0.371±0.022	**0.439±0.038**
Magic K: 0.648	Freq	0.409±0.018	0.182±0.013	0.194±0.012	**0.423±0.010**	0.168±0.014	0.312±0.014	0.304±0.015	0.223±0.023	0.156±0.018
	Clf	0.453±0.021	0.485±0.015	0.488±0.013	**0.557±0.011**	0.440±0.021	0.543±0.012	0.534±0.011	0.204±0.060	0.414±0.009
RT1 K: 0.454	Freq	**0.747±0.002**	0.301±0.047	0.290±0.059	0.596±0.153	0.289±0.058	0.567±0.089	0.555±0.106	−0.172±0.059	0.464±0.036
	Clf	**0.652±0.006**	0.212±0.037	0.203±0.044	0.505±0.135	0.201±0.044	0.450±0.067	0.438±0.083	−0.207±0.048	0.444±0.032
RT2 K: 0.481	Freq	**0.613±0.016**	−0.015±0.176	0.068±0.138	0.574±0.081	−0.072±0.164	0.308±0.167	0.301±0.175	−0.005±0.105	0.397±0.125
	Clf	**0.485±0.015**	−0.023±0.147	0.047±0.117	0.474±0.055	−0.065±0.138	0.245±0.131	0.242±0.135	−0.093±0.082	0.394±0.107
RBF1 K: 0.400	Freq	**0.436±0.018**	−0.146±0.019	−0.110±0.012	0.205±0.017	−0.145±0.014	−0.072±0.017	−0.069±0.016	−0.371±0.017	0.411±0.013
	Clf	0.199±0.031	−0.186±0.029	−0.163±0.024	0.186±0.034	−0.180±0.025	−0.153±0.026	−0.156±0.019	**−0.273±0.017**	0.272±0.023
RBF2 K: 0.411	Freq	**0.431±0.013**	−0.222±0.013	−0.186±0.013	0.122±0.019	−0.219±0.013	−0.161±0.014	−0.167±0.015	−0.296±0.021	0.355±0.019
	Clf	0.298±0.008	−0.266±0.013	−0.271±0.014	0.257±0.035	−0.254±0.018	−0.272±0.012	−0.276±0.014	−0.395±0.014	**0.406±0.015**
RT1 Imb K: 0.427	Freq	0.827±0.002	0.530±0.012	0.527±0.012	**0.918±0.008**	0.527±0.011	0.909±0.008	0.909±0.008	−0.560±0.053	0.634±0.028
	Clf	0.606±0.001	0.398±0.009	0.397±0.009	**0.657±0.006**	0.395±0.009	**0.657±0.006**	0.656±0.005	−0.371±0.043	0.428±0.027
RT2 Imb K: 0.581	Freq	0.883±0.001	0.376±0.033	0.376±0.033	**0.886±0.006**	0.359±0.032	0.879±0.005	0.879±0.006	−0.053±0.074	0.366±0.042
	Clf	0.646±0.001	0.303±0.020	0.303±0.020	0.649±0.004	0.291±0.020	**0.650±0.004**	**0.650±0.004**	−0.048±0.054	0.278±0.034
RBF1 Imb K: 0.374	Freq	**0.904±0.002**	0.517±0.024	0.518±0.024	0.871±0.018	0.506±0.023	0.867±0.014	0.866±0.015	−0.616±0.045	0.657±0.035
	Clf	0.639±0.001	0.426±0.012	0.427±0.012	0.688±0.003	0.417±0.011	**0.676±0.002**	**0.676±0.002**	−0.575±0.033	0.598±0.032
RBF2 Imb K: 0.477	Freq	**0.934±0.002**	0.517±0.001	0.518±0.002	0.926±0.003	0.492±0.007	0.915±0.003	0.915±0.003	−0.456±0.089	0.565±0.065
	Clf	0.692±0.001	0.406±0.004	0.407±0.004	0.710±0.002	0.387±0.006	**0.700±0.001**	**0.700±0.001**	−0.612±0.054	0.683±0.039

the highest correlation with $Freq$ and Clf for the Balance dataset instead. This shows that the difficult to classify instances for the Balance dataset are instances that experience many label fluctuations. For the Codon dataset, the difficult to classify instances require a stronger ensemble model with more weak learners or more data to be correctly classified. In this case, instances from the Balance dataset are difficult due to high correctness difficulty, and instances from the Codon dataset are difficult due high label difficulty.

Given that some of the proposed difficulty measures are based on the label fluctuations (F, F_{cnt}, F_{peak}, A_{max}, A_{sum} and IF), these difficulty measures behave differently for datasets that have different levels of classification difficulty. An example of difficulty to classify dataset is the Abalone dataset, where the kappa performance of the final ensemble model is only 0.117. This suggests that the instances within the dataset cannot form a strong underlying concept that can correctly classify the instances. For these datasets, the label fluctuation-based difficulty measures are often inversely correlated with the difficult to classify instances. The easy to classify instances in the dataset are instances with more label fluctuations. This inverse correlation happens because label fluctuation-based difficulty measures are correctness difficulty and only look at fluctuations in the label. For datasets with weak underlying concepts, most instances in the dataset are incorrectly labeled for an extended number of iterations. In this scenario, these difficult to classify instances will have no fluctuations in the label. The easier to classify instances will be the ones that can occasionally be correctly classified throughout the ensemble process. The

Table 2. The correlations of proposed difficulty measures with data complexity measures dNN and CB.

Dataset	Measure	C	F_{cnt}	F	F_{max}	F_{peak}	A_{max}	A_{sum}	IF	IE
Iris K: 0.916	dNN	0.566 ± 0.140	0.546 ± 0.119	0.610 ± 0.049	0.592 ± 0.131	0.308 ± 0.086	$\mathbf{0.656\pm0.051}$	0.636 ± 0.051	0.129 ± 0.136	0.229 ± 0.066
	CB	NA	NA	NA	NA	NA	NA	NA	NA	NA
Ecoli K: 0.719	dNN	0.233 ± 0.030	0.346 ± 0.035	0.335 ± 0.020	$\mathbf{0.473\pm0.027}$	0.337 ± 0.036	0.347 ± 0.028	0.284 ± 0.029	-0.244 ± 0.050	0.246 ± 0.060
	CB	$\mathbf{0.799\pm0.051}$	0.394 ± 0.035	0.488 ± 0.030	0.546 ± 0.042	0.321 ± 0.040	0.568 ± 0.036	0.579 ± 0.040	-0.119 ± 0.080	0.150 ± 0.095
Car K: 0.678	dNN	0.359 ± 0.000	0.412 ± 0.002	0.420 ± 0.002	0.430 ± 0.003	0.367 ± 0.010	0.435 ± 0.002	0.439 ± 0.002	-0.114 ± 0.056	$\mathbf{0.683\pm0.001}$
	CB	$\mathbf{0.991\pm0.000}$	0.628 ± 0.022	0.670 ± 0.019	0.666 ± 0.011	0.556 ± 0.022	0.720 ± 0.012	0.770 ± 0.009	0.232 ± 0.074	0.672 ± 0.006
Spambase K: 0.875	dNN	0.201 ± 0.006	0.168 ± 0.005	0.165 ± 0.005	$\mathbf{0.235\pm0.004}$	0.170 ± 0.007	0.204 ± 0.005	0.193 ± 0.005	0.134 ± 0.013	0.224 ± 0.007
	CB	0.197 ± 0.040	0.163 ± 0.025	0.165 ± 0.031	0.185 ± 0.024	0.116 ± 0.007	0.196 ± 0.032	0.210 ± 0.037	0.188 ± 0.024	-0.226 ± 0.039
Yeast K: 0.316	dNN	0.212 ± 0.008	-0.144 ± 0.007	-0.176 ± 0.006	0.131 ± 0.028	-0.142 ± 0.008	-0.192 ± 0.012	-0.213 ± 0.012	-0.183 ± 0.009	$\mathbf{0.225\pm0.008}$
	CB	0.105 ± 0.036	$\mathbf{-0.534\pm0.006}$	-0.356 ± 0.017	-0.236 ± 0.030	-0.475 ± 0.014	-0.249 ± 0.032	-0.170 ± 0.032	-0.234 ± 0.069	0.202 ± 0.062
Abalone K: 0.117	dNN	0.265 ± 0.009	-0.299 ± 0.009	-0.301 ± 0.009	0.007 ± 0.020	-0.291 ± 0.010	-0.292 ± 0.008	-0.204 ± 0.015	0.230 ± 0.013
	CB	0.381 ± 0.032	$\mathbf{-0.481\pm0.028}$	-0.441 ± 0.024	-0.110 ± 0.034	-0.459 ± 0.021	-0.416 ± 0.022	-0.420 ± 0.023	-0.135 ± 0.050	0.182 ± 0.043
Nursery K: 0.635	dNN	-0.021 ± 0.000	0.016 ± 0.000	0.016 ± 0.000	-0.013 ± 0.000	0.016 ± 0.000	0.011 ± 0.000	0.011 ± 0.000	$\mathbf{0.022\pm0.000}$	-0.022 ± 0.000
	CB	$\mathbf{0.782\pm0.011}$	0.624 ± 0.040	0.358 ± 0.047	0.383 ± 0.048	0.596 ± 0.043	0.339 ± 0.047	0.428 ± 0.042	-0.114 ± 0.104	-0.012 ± 0.077
Adult K: 0.576	dNN	0.500 ± 0.016	0.370 ± 0.026	0.370 ± 0.023	$\mathbf{0.513\pm0.013}$	0.273 ± 0.028	0.462 ± 0.010	0.457 ± 0.011	0.299 ± 0.017	0.422 ± 0.041
	CB	$\mathbf{0.716\pm0.049}$	0.446 ± 0.024	0.523 ± 0.004	0.653 ± 0.039	0.147 ± 0.003	0.686 ± 0.028	0.702 ± 0.030	0.089 ± 0.016	0.424 ± 0.097
Magic K: 0.648	dNN	0.412 ± 0.027	0.308 ± 0.006	0.311 ± 0.006	$\mathbf{0.491\pm0.005}$	0.279 ± 0.009	0.411 ± 0.004	0.405 ± 0.004	0.200 ± 0.041	0.447 ± 0.013
	CB	0.198 ± 0.076	0.201 ± 0.019	0.187 ± 0.017	0.367 ± 0.022	0.189 ± 0.022	0.281 ± 0.016	0.274 ± 0.017	-0.194 ± 0.022	$\mathbf{0.633\pm0.019}$
RT1 K: 0.454	dNN	$\mathbf{0.416\pm0.006}$	0.093 ± 0.022	0.087 ± 0.027	0.314 ± 0.087	0.086 ± 0.027	0.256 ± 0.045	0.250 ± 0.053	-0.163 ± 0.040	0.305 ± 0.025
	CB	$\mathbf{0.886\pm0.007}$	0.324 ± 0.058	0.310 ± 0.070	0.683 ± 0.183	0.311 ± 0.069	0.636 ± 0.102	0.620 ± 0.124	-0.227 ± 0.058	0.561 ± 0.042
RT2 K: 0.481	dNN	$\mathbf{0.365\pm0.012}$	-0.046 ± 0.089	-0.003 ± 0.070	0.334 ± 0.036	-0.073 ± 0.083	0.143 ± 0.088	0.144 ± 0.091	-0.071 ± 0.052	0.292 ± 0.074
	CB	$\mathbf{0.761\pm0.018}$	-0.048 ± 0.173	-0.048 ± 0.135	0.657 ± 0.063	-0.141 ± 0.154	0.357 ± 0.164	0.358 ± 0.178	0.051 ± 0.108	0.467 ± 0.130
RBF1 K: 0.400	dNN	0.032 ± 0.042	$\mathbf{-0.189\pm0.032}$	-0.124 ± 0.037	0.033 ± 0.055	-0.182 ± 0.024	-0.096 ± 0.036	-0.089 ± 0.031	-0.052 ± 0.018	0.033 ± 0.020
	CB	$\mathbf{0.570\pm0.016}$	-0.138 ± 0.019	-0.124 ± 0.009	0.094 ± 0.033	-0.116 ± 0.018	-0.084 ± 0.017	-0.082 ± 0.011	-0.274 ± 0.037	0.325 ± 0.029
RBF2 K: 0.411	dNN	0.236 ± 0.025	-0.198 ± 0.017	-0.139 ± 0.012	0.083 ± 0.025	-0.208 ± 0.013	-0.124 ± 0.010	-0.118 ± 0.011	-0.193 ± 0.016	0.223 ± 0.017
	CB	$\mathbf{0.700\pm0.016}$	-0.282 ± 0.015	-0.261 ± 0.012	0.127 ± 0.028	-0.242 ± 0.016	-0.227 ± 0.011	-0.236 ± 0.007	-0.247 ± 0.021	0.338 ± 0.014
RT1 Imb K: 0.427	dNN	0.584 ± 0.000	0.326 ± 0.007	0.325 ± 0.007	$\mathbf{0.611\pm0.007}$	0.325 ± 0.008	0.602 ± 0.005	0.601 ± 0.005	-0.352 ± 0.043	0.405 ± 0.027
	CB	0.857 ± 0.000	0.507 ± 0.011	0.503 ± 0.011	$\mathbf{0.920\pm0.009}$	0.504 ± 0.011	0.907 ± 0.008	0.907 ± 0.008	-0.554 ± 0.055	0.630 ± 0.029
RT2 Imb K: 0.581	dNN	0.599 ± 0.000	0.254 ± 0.017	0.254 ± 0.017	0.598 ± 0.001	0.242 ± 0.017	0.594 ± 0.001	0.594 ± 0.001	-0.015 ± 0.054	0.231 ± 0.030
	CB	0.911 ± 0.000	0.394 ± 0.032	0.394 ± 0.032	$\mathbf{0.912\pm0.004}$	0.378 ± 0.031	0.907 ± 0.004	0.907 ± 0.004	-0.051 ± 0.074	0.374 ± 0.042
RBF1 Imb K: 0.374	dNN	-0.001 ± 0.002	0.037 ± 0.016	0.039 ± 0.014	0.103 ± 0.016	0.032 ± 0.019	0.074 ± 0.010	0.075 ± 0.010	-0.208 ± 0.029	0.195 ± 0.028
	CB	$\mathbf{0.982\pm0.001}$	0.476 ± 0.030	0.474 ± 0.029	0.864 ± 0.020	0.471 ± 0.028	0.862 ± 0.019	0.862 ± 0.019	-0.576 ± 0.047	0.624 ± 0.036
RBF2 Imb K: 0.477	dNN	0.339 ± 0.001	0.188 ± 0.005	0.191 ± 0.004	$\mathbf{0.368\pm0.003}$	0.175 ± 0.006	0.355 ± 0.001	0.355 ± 0.001	-0.110 ± 0.069	0.145 ± 0.057
	CB	$\mathbf{0.986\pm0.000}$	0.500 ± 0.008	0.499 ± 0.008	0.948 ± 0.002	0.480 ± 0.011	0.931 ± 0.002	0.932 ± 0.002	-0.448 ± 0.088	0.560 ± 0.062

negative correlation with correctness difficulty measures shows that the dataset has a high label difficulty, and the instances required more relevant data to be correctly labeled.

(RQ 3). The correlations of the nine proposed difficulty metrics with the two data complexity measures dNN and CB that measures class overlap and class imbalance respectively are shown in Table 2. The correlation with class balance for datasets Iris and Segment are "NA" because these two datasets are balanced, with an equal number of instances per class; therefore, all instances will have the same CB value. Similar to the correlation results with $Freq$ and Clf, none of the difficulty measures have the strongest correlation with dNN or CB for all datasets. In general, there is a strong correlation (>0.1) between the difficulty measures and the two data complexity measures, with the difficulty measures having a stronger correlation with CB in comparison to dNN. This indicates that difficulty measures are able to identify both instances from minority classes and borderline points as instances that are difficult to classify. The fluctuation-based difficulty measures are also inversely correlated to both dNN and CB for datasets with low performance scores. Same as the previous correlation results with the uncertainty-based difficulty measure, we use the Abalone dataset as an example. For this dataset, the fluctuation score F has the strongest correlation with dNN with a value of -0.301, and the fluctuation count F_{cnt} has the strongest correlation with CB with a value of -0.481. This shows that these instances with high dNN and CB values do not have many label fluctuations and require additional data to be correctly classified.

5 Conclusion

We introduced nine difficulty measures that discover difficult instances based on a boosted ensemble model. The boosting model is constructed through an additive process with an increasing number of weak learners per iteration. We monitor the number of learners required and the label fluctuations as indications of classification difficulty at the instance level. We also distinguish between two types of classification difficulty: label difficulty and correctness difficulty. In addition, we show the correlations between the difficulty measures and also in relation to existing difficulty measures. The results show that there is a correlation between the proposed difficulty measures and existing uncertainty measures, as well as a significant correlation with class balance and neighborhood data complexity measures.

References

1. Abe, N., Zadrozny, B., Langford, J.: Outlier detection by active learning. In: Proceedings of the 12th ACM SIGKDD International Conference on Knowledge Discovery and Data Mining, pp. 504–509 (2006)
2. Anwar, N., Jones, G., Ganesh, S.: Measurement of data complexity for classification problems with unbalanced data. Stat. Anal. Data Min. ASA Data Sci. J. **7**(3), 194–211 (2014)
3. Armano, G., Tamponi, E.: Experimenting multiresolution analysis for identifying regions of different classification complexity. Pattern Anal. Appl. **19**(1), 129–137 (2016). https://doi.org/10.1007/s10044-014-0446-y
4. Chen, T., Guestrin, C.: XGBoost: a scalable tree boosting system. In: Proceedings of the 22nd ACM SIGKDD International Conference on Knowledge Discovery and Data Mining (2016)
5. Dietterich, T.G.: An experimental comparison of three methods for constructing ensembles of decision trees: bagging, boosting, and randomization. Mach. Learn. **40**(2), 139–157 (2000). https://doi.org/10.1023/A:1007607513941
6. Fernández, A., García, S., Galar, M., Prati, R.C., Krawczyk, B., Herrera, F.: Learning from Imbalanced Data Sets, vol. 10. Springer, Cham (2018). https://doi.org/10.1007/978-3-319-98074-4
7. Freund, Y., Schapire, R.E.: A decision-theoretic generalization of on-line learning and an application to boosting. J. Comput. Syst. Sci. **55**(1), 119–139 (1997)
8. Friedman, J.H.: Stochastic gradient boosting. Comput. Stat. Data Anal. **38**, 367–378 (2002)
9. Garcia, L.P., de Carvalho, A.C., Lorena, A.C.: Effect of label noise in the complexity of classification problems. Neurocomputing **160**, 108–119 (2015)
10. Hastie, T., Rosset, S., Zhu, J., Zou, H.: Multi-class AdaBoost. Stat. Interface **2**(3), 349–360 (2009)
11. Ho, T.K., Basu, M.: Complexity measures of supervised classification problems. IEEE Trans. Pattern Anal. Mach. Intell. **24**(3), 289–300 (2002)
12. Hüllermeier, E., Waegeman, W.: Aleatoric and epistemic uncertainty in machine learning: an introduction to concepts and methods. Mach. Learn. **110**(3), 457–506 (2021)

13. Leyva, E., González, A., Perez, R.: A set of complexity measures designed for applying meta-learning to instance selection. IEEE Trans. Knowl. Data Eng. **27**(2), 354–367 (2014)
14. Lorena, A.C., Costa, I.G., Spolaôr, N., De Souto, M.C.: Analysis of complexity indices for classification problems: Cancer gene expression data. Neurocomputing **75**(1), 33–42 (2012)
15. Montiel, J., Read, J., Bifet, A., Abdessalem, T.: Scikit-multiflow: a multi-output streaming framework. J. Mach. Learn. Res. **19**(72), 1–5 (2018). http://jmlr.org/papers/v19/18-251.html
16. Nguyen, V.L., Shaker, M.H., Hüllermeier, E.: How to measure uncertainty in uncertainty sampling for active learning. Mach. Learn. **111**(1), 89–122 (2022). https://doi.org/10.1007/s10994-021-06003-9
17. Nussbaumer, H.J.: The fast Fourier transform. In: Nussbaumer, H.J. (ed.) Fast Fourier Transform and Convolution Algorithms. Springer Series in Information Sciences, vol. 2, pp. 80–111. Springer, Heidelberg (1981). https://doi.org/10.1007/978-3-662-00551-4_4
18. Pungpapong, V., Kanawattanachai, P.: The impact of data-complexity and team characteristics on performance in the classification model. Int. J. Bus. Anal. **9**, 1–16 (2022)
19. Schober, P., Boer, C., Schwarte, L.A.: Correlation coefficients: appropriate use and interpretation. Anesth. Analg. **126**(5), 1763–1768 (2018)
20. Smith, M.R., Martinez, T., Giraud-Carrier, C.: An instance level analysis of data complexity. Mach. Learn. **95**(2), 225–256 (2014). https://doi.org/10.1007/s10994-013-5422-z
21. Wang, H., Bah, M.J., Hammad, M.: Progress in outlier detection techniques: a survey. IEEE Access **7**, 107964–108000 (2019)

Graph Embeddings for Non-IID Data Feature Representation Learning

Qiang Sun[1,2](✉) ⓘ, Wei Liu[1,2] ⓘ, Du Huynh[1] ⓘ, and Mark Reynolds[1] ⓘ

[1] UWA NLP-TLP Group, Perth, Australia
pascal.sun@research.uwa.edu.au,
{wei.liu,du.huynh,mark.reynolds}@uwa.edu.au
[2] School of Physics, Mathematics and Computing, The University of Western Australia, Perth, Australia
https://nlp-tlp.org/

Abstract. Most machine learning models like Random Forest (RF) and Support Vector Machine (SVM) assume that features in the datasets are independent and identically distributed (IID). However, many datasets in the real world contain structural dependencies so neither the data observations nor the features satisfy this IID assumption. In this paper, we propose to incorporate the latent structural information in the data and learn the best embeddings for the downstream classification tasks. Specifically, we build traffic knowledge graphs for a traffic-related dataset and apply node2vec and TransE to learn the graph embeddings, which are then fed into three machine learning algorithms, namely SVM, RF, and kNN to evaluate their performance on various classification tasks. We compare the performance of these three classification models under two different representations of the same dataset: the first representation is based on traffic speed, volume, and speed limit; the second representation is the graph embeddings learned from the traffic knowledge graph. Our experimental results show that the road network information captured in the knowledge graphs is crucial for predicting traffic risk levels. Through our empirical analysis, we demonstrate knowledge graphs can be effectively used to capture the structural information in no-IID datasets.

Keywords: Graph embedding · Knowledge graph · Non-IID

1 Introduction

When we talk about data analysis and machine learning, we mainly refer to 'tidy data' in tabular format. A row in a table represents an entity together with its properties. Since we live in a connected world, these entities within a table are usually related. However, a tabular representation of data cannot express these interrelationships effectively. A *graph* is a simple structure containing nodes and edges, which represents not only information about entities, but also interrelationships between entities. *Knowledge graphs* (KGs) are a special type of graphs

© The Author(s), under exclusive license to Springer Nature Singapore Pte Ltd. 2022
L. A. F. Park et al. (Eds.): AusDM 2022, CCIS 1741, pp. 43–57, 2022.
https://doi.org/10.1007/978-981-19-8746-5_4

where entities and relations are of different types, and are used to express the real-world knowledge.

With the development of Cloud Computing, Big Data, and the Internet of Things (IoT), many large-scale industrial KGs can then be established from real-world data in addition to the general-purpose KGs mentioned above. However, it is hard for machines to process and analyse large-scale KGs efficiently and effectively, which makes it difficult for human to navigate to meet analytical needs.

Knowledge graph embeddings (KGEs) [1] provide an effective and efficient solution. They learn the low dimensional representations of large-scale KGs while preserving the graph structural information. Downstream machine learning tasks, such as link prediction and node classification, can then be applied to the lower dimensional space. With the reduced time and space cost, human beings can explore, interpret, gain insights and benefit from the large-scale KGs more easily. It brings more possibility.

A fair amount of knowledge graph embeddings (KGE) models have been proposed in the last decade [1], leading to several practical applications. For example, TransE [2] helps to improve the efficiency of recommender systems due to its good performance in the link prediction area. However, most of the current research community focuses on analysing general-purpose knowledge graphs. Suppose we want to apply these findings to industries, there are generally two steps to follow: the first step is to construct a knowledge graph based on the data; the second step is to apply proper knowledge graph embedding models to learn the embeddings and evaluate their performances to yield the best model. Most of the current research either focus on the former (how to build a knowledge graph) or the latter (how to analyse a general knowledge graph), with very little on combining the two to verify the effectiveness of these models when applied on real-world data from industries.

The contribution of this paper is to demonstrate how representing Non-IID data using knowledge graphs and learning feature representations with node2vec [3] can improve the accuracy of classification tasks.

2   Background and Related Work

2.1   Classification Models and IID Assumption

Traditionally, when we deal with a classification task, we first collect data and represent all observable properties or selected properties we think are relevant to the predicted target as features, then use various machine learning models. In machine learning, several classical algorithms excel in dealing with classification problems. These include random forests (RF), k-Nearest Neighbours (kNN), support vector machines (SVM), logistic regression and Naïve Bayes. Given the variables X, Logistics regression calculates the probability of the dependant target variable Y and then gives a binary output. The Naïve Bayes classifier, based on Bayes' theorem, assumes all feature variables in the dataset are independent when estimating the probability of whether a test instance belongs to a particular category. Both logistic regression and Naïve Bayes hold the independent and identically distributed (IID) [4] assumption for observations.

Decision trees (DT) are constructed by splitting datasets via answering a set of questions. DT holds the IID assumption [5]. Random forest (RF), like its name implies, is composed of several decision trees. Random forest is an ensemble method, so the essence of the principle still comes from the decision trees. In addition, RF requires random re-sampling, so it also holds the IID assumption [5] for observations.

The k-Nearest Neighbours (kNN) is an instance-based algorithm. kNN considers the similarity between instances. In other words, it considers the relative virtual position of each entities. kNN does not make any assumptions about the features, which indicates it does not hold IID assumption [6] for observations. The assumptions each models have made with observations have been summarised in Table 1.

Table 1. Traditional classification models

Models	Observations
Logistic Regression	Independence assumed
Naïve Bayes	Independence assumed
SVM	Independence assumed
DT/RF	Independence assumed
kNN	No assumption

While using these methods, we assume that the observations represented by each row of data in the table are relatively independent and identical distributed. That is, the classification of each entity depends on the nature of that entity itself and is not affected by other entities. It is a reasonable assumption for tasks like spam classification and works effectively.

For datasets where entities or data instances have interrelationships, the relationships between entities will lead to connections between the individual features. Traffic-related datasets, for example, which focus on road segments as observations, cannot be sufficiently stored as tables, because road networks are naturally connected with structural dependencies that need to be captured. Therefore, theoretically, of these five types of models, kNN should be the best performer. However, we also aim to take advantage of structural information with datasets while kNN only holds no IID assumption with datasets. So, we need to explicitly represent the relationships between entities via graphs, we should be able to achieve better performance in classification tasks. If we represent the network information with a graph. The whole road networks can be divided into multiple connected road segments. Each node in graphs represent a road segment, and edges represent how road segments connect. In this way, the relative spatial information between road segments is well presented in the simple traffic knowledge graph.

After the graph construction step, the next challenge is to interpret the graph. We cannot perform the classification task directly with the graph. Ideally, we want to process the graph, retain the structural information, and present the results in tabular format to traditional classification models. Embedding is one of the ways to achieve this purpose.

2.2 Graph and Knowledge Graph Embeddings

Embedding originates from word2vec [7] in the natural language process-ing (NLP) field, transforming a high-dimensional vector into a low-dimensional vector representation. Since then, the concept of embedding has been applied to various areas of data analysis with significant impact. Node2Vec [3] is one of the methodologies and was introduced in 2016. It is inspired by deepwalk [8], anal-ysed Breadth-first Sampling (BFS) and Depth-first Sampling (DFS) algorithms, found that BFS can help obtain structural equivalence. At the same time, DFS has the advantage of obtaining homophily. They proposed a probability-based biased random walk strategy to interpolate between BFS and DFS. The gener-ated nodes sequences are then feed into word2vec [7] to obtain node embeddings. The probability formula [3] is shown in Eq. 1, where Z is a normalisation con-stant, $d(u, v)$ returns the distance between two nodes u and v. p and q are the hyper-parameters that change the behaviour of the walk to behave more like BFS or DFS. Here, p adjusts the probability for revisiting the previous node, is called *return parameter*. A lower value of p means search depth in DFS is shal-lower and fewer neighbours in BFS. Lower q, which is called $in-outparameter$, leads to more DFS-like search. While higher q leads to more BFS-like search. Another 3 hyper-parameters include: number of walks r, maximum walk length l and embedding dimension D

$$\Pr(v_{i+1}|v_i, v_{i-1}) = \frac{1}{Z} \cdot \begin{cases} \frac{1}{p} & \text{if } d(v_{i-1}, v_{i+1}) = 0 \\ 1 & \text{if } d(v_{i-1}, v_{i+1}) = 1 \\ \frac{1}{q} & \text{if } d(v_{i-1}, v_{i+1}) = 2 \\ 0 & \text{if } d(v_{i-1}, v_{i+1}) \notin E \end{cases} \tag{1}$$

Node2vec can be seen as an upgraded version of deepwalk [8] and a cus-tomised version of word2vec [7], a deep learning based graph embedding app-roach. Furthermore, it can only work effectively on weighted or unweighted homogeneous graphs, because it can only process graphs with one type of node and edge respectively. Our road network knowledge graph is a homogeneous graph with only **RoadSegment** node type, so we can use node2vec [3] to obtain road network graphs embeddings.

TransE [2] is also inspired by word2vec [7]. The word2vec [7] embeds words into vectors. The translation invariance (shown in Eq. (2) [1]) found in the word embedding space inspired translation distance-based graph embedding models.

$$\overrightarrow{King} - \overrightarrow{Queen} = \overrightarrow{Man} - \overrightarrow{Woman} \tag{2}$$

where $\vec{W}$ is the embedding vector for word W. TransE [2] is the first piece of work in translation distance-based model series for knowledge graph embeddings. It can not only convert knowledge graph nodes and edges into low-dimension embeddings, but also help us better interpret the relationships between nodes.

2.3 Summary

Our plan is first to train classification models with the traditional classification algorithms. Then we use the same datasets to build a road network graph and a simple traffic knowledge graph, apply node2vec [3] and TransE [2] against the graph, respectively. Then we will feed the embeddings into traditional classification models to fulfil the same classification tasks. In the end, we will evaluate the results to determine which the performance of methodologies.

3 Methodology

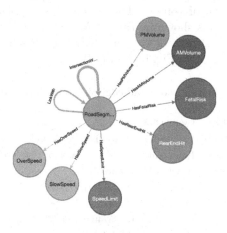

Fig. 1. Traffic knowledge graph schema

A **knowledge graph** can have multiple types of nodes and edges, so to build a knowledge graph, we need to first define the node types, nodes and edge types. We can add attributes to nodes and can use types to represent some extra information in knowledge graph, however, the TransE [2] model will not take that into account.

We selected 9 types of edges and 8 types of nodes to build a simple knowledge graph for traffic information as shown in Fig. 1. Node **RoadSegment** is defined as a unique part of the road networks. Relative spatial relationships between road segments can be reflected in the graph via connectivity of the **RoadSegment**. An edge is established when a **RoadSegment** is connected with another.

Road segments are connected in both directions, so the graph we constructed is undirected.

Then we add more types of nodes by promoting attributes of road segments into multiple nodes. We define two types of nodes and edges regarding risk. One type is **FatalRisk** node, which will include 3 nodes, **High/Medium/None** risk nodes. Road segment labelled as high risk will link to **FatalRisk (High)** node via edge **HasFatalRisk**. Another node type regarding risk is **RearEndHit**. If a road segment had been involved in a rear-end crash, then this road segment will link with node *RearEndHit* via **hasRearEndHit**. Node type **SpeedLimit** contains 6 nodes, road segment links with their speed limit via edge **Has-SpeedLimit**. Depending on whether field **85th percentile Speed > Speed Limit** value is greater than 0 or less than 0, we define two types of nodes: **Over-Speed** and **SlowSpeed**. Further, based on the value of **85th percentile Speed > Speed Limit**, we classify **OverSpeed** into 5 categories: $[O_1, O_2, O_3, O_4, O_5]$, and **SlowSpeed** into 5 categories: $[L_1, L_2, L_3, L_4, L_5]$, each difference of 5 km/h is a classification. We use node type **AMVolume** and **PMVolume** to characterise the traffic volume information in this traffic knowledge graph. Based on traffic volumes, each node type will have 4 nodes: **H/M/L/N**. Each difference of 200 is a classification of volume. The summary statistics about this simple traffic knowledge graph is shown in Table 2. We use Neo4j[1] to build the knowledge graph, and visualisation for the knowledge graph schema and part of the knowledge graph are show in Fig. 1 and Fig. 2.

Table 2. Types of nodes and edges in traffic knowledge graph

Name	Type	Comments	Numbers
RoadSegment	Node	Use ASSET_ID to distinguish nodes	2287
FatalRisk	Node	Enumerate, H/M/N, stands for risk level	3
RearEndHit	Node	RearEndHit occurred	1
OverSpeed	Node	Enumerate, $O_1/O_2/O_3/O_4/O_5$	5
SlowSpeed	Node	Enumerate, $L_1/L_2/L_3/L_4/L_5$	5
SpeedLimit	Node	Enumerate, 15/25/40/50/60/70 (km/h)	6
AMVolume	Node	Enumerate, H/M/L/N	4
PMVolume	Node	Enumerate, H/M/L/N	4
LinkWith	Edge	Link connected RoadSegment	1086
IntersectionWith	Edge	The link between road segments is intersection	2686
HasFatalRisk	Edge	Link RoadSegment with FatalRisk nodes	2287
HasRearEndHit	Edge	Link RearEndHit with RoadSegment	117
HasOverSpeed	Edge	Link OverSpeed nodes with RoadSegment	423
HasSlowSpeed	Edge	Link SlowSpeed nodes with RoadSegment	548
HasSpeedLimit	Edge	Link SpeedLimit nodes with RoadSegment	2287
HasAMVolume	Edge	Link AMVolume nodes with RoadSegment	974
HasPMVolume	Edge	Link PMVolume nodes with RoadSegment	974

---

[1] https://neo4j.com/.

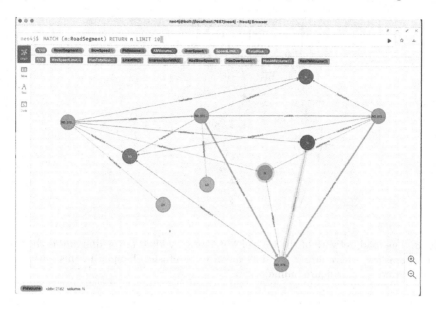

Fig. 2. Part of the simple traffic knowledge graph

4   Dataset and Experiment Design

4.1   Dataset

Our traffic dataset[2] was provided by City of Mitcham, South Australia. The road network is divided into multiple road segments. Each road segment has its unique id. Traffic engineers are mainly concerned about the AM (7–9 am) and PM (4–6pm) peak hour traffic volumes, as most traffic congestion are caused by the peak hour traffic. The AM and PM peak hour traffic volumes for both traffic flow directions are provided in the dataset.

According to *AustRoads*, the definition of 85th percentile speed [9] is: "The speed at or below which 85% of all vehicles are observed to travel under free-flowing conditions past a monitored point." For local councils, speed limits include 40 km/h, 50 km/h, 60 km/h and 70 km/h. As the 85th percentile speed is used by traffic engineers as a guide to set a safe speed limit, the differences between speed limits and the 85th percentile speeds are also provided in the dataset for both traffic directions.

Crashes since 2016 are also stored in the dataset. The data for each crash includes: *location, crash severity* (divided into *fatality, injury,* and *property damage only*), and *crash type* (e.g., *right angle, hit a fixed object,* or *rear-end hit*). The original crash data came from the *South Australia Viewer* database. We prepossessed the dataset so that the crashes are associated with proper road segments.

---

[2] The dataset url: https://wgaz.maps.arcgis.com/apps/webappviewer/index.html? id=71836e2da20e435c888d5b5e7e72c3ee.

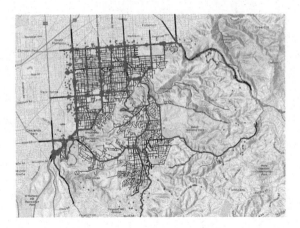

Fig. 3. The road network of the City of Mitcham. Overlaid in the diagram is the heat-map of crashes, where larger red dots mean more crashes happen in this part of the road network. (Color figure online)

We also prepossessed the dataset to remove duplication based on *ASSET_ID* field. The entire road network, shown in Fig. 3, consists of 2,287 road segments. Each road segment has 216 features, including road names, suburbs, directions (north-south or east-west), pavement conditions, speed limits, etc. Due to practical reasons, there are missing data in the dataset. For example, it is not practical to conduct the surveys across all the road segments within a five year time frame.

If the road segment is located on the main road of the traffic network, this means higher traffic volume and speed, often associated with more accidents. Traffic volumes, speeds, and risks of road segments are correlated, they are also influenced by the road network structure. Traditionally, volume, speed, and crash information are treated as independent features, as implemented in WAS, and the connection information among road segments has not been considered.

If we select all the features for model training, the difficulty and workload of data prepossessing will be very high. At the same time, it will not be conducive to our subsequent model evaluation. As the main areas of concern are speed, traffic volume, and road safety, for traffic speeds, we decided to use *speed limit* and *85th percentile Speed − Speed Limit* to represent the speed information; for the traffic flows, we used only the AM and PM peak hour traffic counts, which best reflect the traffic congestion situation.

For traffic crashes (see Fig. 3), we decided to select the *TOTAL CASUAL-TIES, TOTAL CRASHES* and *rear-end hit* features for further risk analysis.

Table 3 shows all the features after prepossessing. The speed limits on most road segments are 50 km/h. The differences between the 85th percentile speeds and speed limits vary from −30 km/h to 20 km/h, exhibiting a normal distribution. Most of the road segments are quiet during the morning and afternoon peak hours. Over 500 road segments were involved in crashes, around 180 of which resulted in injury or death. At the same time, around 100 of these road segments

Table 3. Selected features from the raw dataset.

Id	Type	Comments
ASSET_ID	String	Unique identity for road segments
SPEED_LIMIT	Integer	40 km/h, 50 km/h, 60 km/h, 70 km/h
85th percentile Speed − Speed Limit	Numeric	Differences value
AM - counts	Integer	AM peak hour volume
PM - counts	Integer	PM peak hour volume
REAR END	Integer	Number of rear-end hit
TOTAL CRASHES	Integer	Total number of crashes
TOTAL CASUALTIES	Integer	Total number of causalities

have been involved in rear-end accidents. There are about 50 road segments on which multiple accidents had occurred.

There is a strong correlation between *REAR END, TOTAL CRASHES*, and *TOTAL CASUALTIES*. This is expected, as rear-end hit and casualties are both included in crashes. There is also a strong correlation between morning and evening peak traffic volumes, while the correlation between speed and traffic volume is weaker.

4.2 Experiment Design

We labelled each road segment by one of the following risk levels: *High*, *Medium*, and *None*, depending on whether a crash had occurred and whether there were casualties. The risk level *None* means no accidents had happened on the road segment before and can therefore be interpreted as in the *low risk* category. These labelled risk levels of road segments were used as the ground truth for the target variable for our downstream risk classification task. We did not take into account the number of crashes in our risk level labelling as only 50 road segments involved multiple crashes.

We first processed the connection information of road segments and turned the road network into a road network graph via Neo4j[3]. In summary, there were 2287 *RoadSegment* nodes and 3772 *Link* edges in the graph. We then applied node2vec [3] to the road network graph and trained it with the following hyper-parameters:

- embedding dimension D, set to 50;
- walk length l, set to 30;
- number of walks r, set to 10;
- return parameter p, set to 1; and
- in-out parameter q, set to 1.

The output embedding for each road segment is a 50-dimensional vector. The above hyper-parameter values were selected based on the experimental results reported in the paper of Jensen et al. [10].

---

[3] https://neo4j.com/.

To train the traffic knowledge graph against the TransE [2], we are using the pykg2vec [11] library, which is an out-of-box software in knowledge graph embedding area. The output dimension for both edges and nodes is 58. We have done the hyper-parameter tuning for this model training process, and the optimised training parameters are: **learning rate** is 0.013, **l1_flag** is true, **hidden_size** is 58, **batch_size** is 116, **epochs** is 200, **margins** is 0.098, **optimiser** is SGD, **sampling** is *bern*, **negative_rate** is 1.

As mentioned above, there are missing data in the dataset. Of the 2287 road segments, 1316 have no speed data, and 1313 have no traffic volume data due to traffic survey limitation. Since the traditional methods cannot handle missing data, we chose to remove the records containing missing data. After this data cleaning step, 969 road segments with no missing data are available. For the traditional methods shown in Table 1, we chose SVM, kNN, and RF for evaluation. These three traditional methods were chosen because they are known to be effective for medium-sized problems. They used the five features shown in the middle part of Table 3 as input features, whereas our proposed graph embedding method is to use the 50-dimensional vector generated from node2vec as input features for these three traditional methods. In other words, there are 9 methods for comparison:

- Using the 5 features shown in Table 3: SVM, kNN, and RF;
- Using the 50-dimensional node2vec embeddings: SVM, kNN, and RF.
- Using the 58-dimensional TransE embeddings: SVM, kNN and RF.

All the methods being compared tackled the same three-class classification problem: predict the *risk levels* of road segments.

Even though all the 2287 road segments had their 50/58-dimensional embedding vectors available, for fair comparison, only those 969 clean road segments were used for all the 9 methods. We divided the 969 clean road segment data into a training set and a test set using the 80/20 split ratio. For each of the three traditional methods using the 5-dimensional input features, we performed hyper-parameter tuning using a validation set extracted from the training set. The following hyper-parameter settings were found to yield the best performing models for comparison:

- for SVM: the **RBF** (radial basis function) kernel was used.
- for kNN: k was set to 5, the distance function *minkowski* was used for the measure of *nearness*;
- for RF: *bootstrap* is set to True, *max_depth* is 3, *max_feature* set to 2, *min_sample_leaf* set to 3, *min_samples_split* is set to 10, and *n_estimators* is 200.

All the methods were trained using the same training set and evaluated on the same test set.

Table 4. Overall accuracy for different input sets and models

Input	Training set			Test set		
	SVM	kNN	RF	SVM	kNN	RF
Traditional_biased	0.688	0.683	0.671	0.686	0.582	0.686
Traditional_unbiased	0.597	0.889	0.771	0.587	0.763	0.665
Node2Vec	1.00	0.992	0.992	0.816	0.777	0.835
TransE	1.000	1.000	0.999	0.845	0.769	0.922

5   Results and Discussions

5.1   Imbalanced Data

As injuries and fatalities are low-probability events, our dataset is highly imbalanced. As a result, the performance of the traditional models trained on the imbalanced dataset was very poor on the test set, in that many road segments for the two under-represented classes (having *High* and *Medium* risks) were incorrectly classified as *None*. Considering the importance of correctly classifying the road segments for the *High* and *Medium* classes, we incorporated:

- oversampling the road segments for the *High* and *Medium* classes to reduce the skewness of the dataset;
- *recall* and *F1 score* for each of the three classes as two additional evaluation measures.

For the traditional methods trained on the original skewed dataset using the 5-dimensional input features, we name them as *traditional_biased*. When they were trained on the balanced dataset, we name them as *traditional_unbiased*. For the traditional methods using the 50-dimensional node2vec embeddings as input, referred as *node2vec*, they were trained on the balanced data only. We will do the same for *TransE*.

5.2   Advantage of Using the Node2vec Embeddings

The accuracy scores of RF and SVM shown in Table 4 are both reasonably good, at around 0.7. However, their recall values for *High* and *Medium* risk levels risks are very poor under the *traditional_biased* rows in Table 5. For the balanced dataset (the *traditional_unbiased* rows), we can see a great improvement on the recall and F1 score values for all the methods. In particular, the severe effect of imbalanced data on kNN is significantly reduced after oversampling [6].

In terms of accuracy after oversampling, the performance of kNN was improved by 18%, beating all three previous models. However, the accuracy of SVM and RF was reduced by 10% and 2% respectively, as shown in Table 4. The most important aspect is that there are significant improvements in recall for both the *Medium* and *High* risk levels for all models, as shown in Table 5. Up

Table 5. Recall and F1 score for each risk level, per input set, per model

Input	Model	Recall			F1 score		
		High	Medium	None	High	Medium	None
traditional_biased	SVM	0.227	0.122	0.992	0.286	0.197	0.836
	kNN	0.136	0.265	0.789	0.146	0.310	0.738
	RF	0.045	0.204	0.992	0.077	0.514	0.602
traditional_unbiased	SVM	0.590	0.470	0.694	0.634	0.514	0.602
	kNN	0.943	0.757	0.587	0.878	0.716	0.683
	RF	0.680	0.496	0.810	0.765	0.567	0.658
node2vec	SVM	1.000	0.791	0.653	0.904	0.795	0.728
	kNN	1.000	0.852	0.479	0.900	0.763	0.617
	RF	0.967	0.809	0.727	0.952	0.788	0.759
TransE	SVM	0.973	0.819	0.739	0.958	0.807	0.762
	kNN	1.000	0.854	0.450	0.912	0.756	0.579
	RF	0.973	0.836	0.950	0.986	0.885	0.891

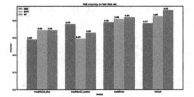 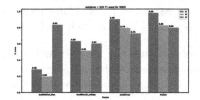

Fig. 4. f1 score and accuracy comparsions between models

until now, kNN has demonstrated to outperform SVM and RF on the balanced dataset while RF is the runner-up, performing slightly better than SVM. These results are expected (Fig. 4).

When we train models using the 50-dimensional vectors from node2vec as input, the performance of all three types of models improves considerably, as shown in Fig. 5, considering our input here is purely road network structure information represented by the 50-dimensional vectors.

In terms of overall accuracy, SVM and RF have improved by 24% and 17%, respectively, both reaching over 80%. kNN accuracy improved by 1% to 0.77. In terms of recall, kNN is overwhelmingly superior in both classifying medium and high risk level among all models. We can tolerate *low risk* being assessed as *medium* or *high* but cannot allow *high risks* to be classified as *medium or low* in this task, thus we will want the sum in the top right corner in Fig. 5 to be as small as possible. kNN wins it with the sum of 9. In conclusion, kNN has the overall best performance when training with node2vec embedding vectors, but the differences with RF and SVM are not as significant as the traditional baseline.

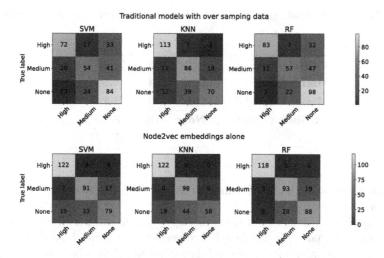

Fig. 5. Confusion matrices for predicting the risk levels from traditional_unbiased and node2vec.

Compared with *node2vec*, TransE [2] embeddings provide extra information about speed, volume and rear-end hit, so the overall performance improve slightly compared with node2vec embeddings as input. In reality, the accuracy of SVM, kNN and RF has improved, with RF improving the most to 0.92, SVM to 0.87 and kNN to 0.78, as we can see from Table 4. The recall of all three risk levels - **High**, **Medium** and **None** has also been improved for all models, according to Table 5.

This result shows that the risk level of each road segment in the traffic network depends mainly on the structure of the entire traffic network and the relative position of the road segment within the network. Speed, traffic and incidents are all manifestations of the network structure. When we take advantage of the spatial information within traffic related datasets, we can achieve better accuracy for classification tasks.

5.3 Evaluation and Discussion

In conclusion, node2vec [3] embeddings as input (*node2vec*) gives better performance compared to results for *traditional_unbiased*. The accuracy of kNN is the best when we feed in the 5 features as input. It becomes the worst when we feed in embeddings as input. It has not gained much improvement when we switch to embeddings as input models. According to Table 1, except for kNN, all other models make the IID assumption. So we can interpret that graph embeddings learn independent identical distributed low-dimensional observations of entities. It helped to greatly improve the accuracy for RF and SVM, but not so much for kNN.

We have 2287 road segments, of which we used 969 for training and testing purposes. TransE embeddings (*TransE*) as input models do not need to con-

Table 6. Risk classification results with TransE + kNN, validation results

Type	Count	Percentage
Correct	903	68.5%
High predicted as **None**	54	4.1%
High predicted as **Medium**	19	1.4%
Medium predicted as **None**	110	8.3%
Other	232	17.7%

sider missing data in the five features so that we can use the rest 1318 rows as validation datasets.

Compared with our ground truth in Fig. 3, the overall performance is good. Road segments that were predicted incorrectly are on the boundary of local council area. The summary of the validation results shown in Table 6 indicates the accuracy is around 68.5%, which is quite impressive, considering we do not hold any extra information with these road segments. This in turn proves that it is vital to leverage the structural information in the dataset. Because in reality, these boundary road segments have connected to road segments from other local government areas. However, this is not represented in our road network graph as it is not in our original datasets.

6   Conclusions and Future Work

6.1   Traffic

The most exciting conclusion from the results is that the accidents are mainly determined by the road network structure. To some extent, road structure determines how the whole traffic network operates, where the congestion happens, which part of roads with high risk.

6.2   Learn Feature Representation for Non-IID Data via Graph Embeddings

When we use a graph to characterise the spatial information of the dataset and then use the low-dimensional representations of the graph learned by node2vec [3] to train the classification models, the models' performance has a relatively significant improvement. So we found that effective use of the spatial information in the dataset can effectively improve the accuracy of the classification task. This is because graph embeddings can learn independent identical distributed low-dimensional observations of entities. Also, *TransE* models perform slightly better than *Node2Vec* models because knowledge graphs hold more information than a simple road network graph. At the same time, because of the natural representation of graphs and knowledge graphs, this methodology can also effectively solve the missing data problem.

6.3 Future Work

We can use road network data from higher administrative levels like Western Australia to validate our conclusions, allowing us to investigate further about the performance, speed, and scalability under larger and more complex Non-IID data. In addition, the dataset contains the length and pavement information of road segments, how to express these information in a graph, and whether the expression of these information will further improve the accuracy is well worth studying.

The traffic knowledge graph we built is heavily skewed, with the nodes of the road segments occupying an absolute majority. We need a more diverse set of datasets containing spatial information to confirm the performance improvement of knowledge graphs for classification tasks and evaluate how much improvement can be achieved.

In general, graph embeddings provide a way to learn lower dimensional representations from Non-IID data, according to our study.

References

1. Dai, Y., Wang, S., Xiong, N., Guo, W.: A survey on knowledge graph embedding: approaches, applications and benchmarks. Electron. (Switz.) **9**, 1–29 (2020)
2. Bordes, A., Usunier, N., Garcia-Duran, A., Weston, J., Yakhnenko, O.: Translating embeddings for modeling multi-relational data. In: Advances in Neural Information Processing Systems, vol. 26 (2013)
3. Grover, A., Leskovec, J.: Node2Vec (2016)
4. O'Neill, B.: Exchangeability, correlation, and Bayes' effect. Int. Stat. Rev. **77**, 241–250 (2009)
5. Gupta, M., Minz, S.: Spatial data classification using decision tree models. In: 2017 Conference on Information and Communication Technology (CICT), pp. 1–5 (2018)
6. Liu, C., Cao, L., Yu, P.: Coupled fuzzy k-nearest neighbors classification of imbalanced non-IID categorical data. In: Proceedings of the International Joint Conference on Neural Networks, pp. 1122–1129 (2014)
7. Mikolov, T., Chen, K., Corrado, G., Dean, J.: Efficient estimation of word representations in vector space. In: 1st International Conference on Learning Representations, ICLR 2013 - Workshop Track Proceedings, pp. 1–12 (2013). https://proceedings.neurips.cc/paper/2013/file/1cecc7a77928ca8133fa24680a88d2f9-Paper.pdf
8. Perozzi, B., Al-Rfou, R., Skiena, S.: DeepWalk: online learning of social representations. In: Proceedings of the ACM SIGKDD International Conference on Knowledge Discovery and Data Mining, pp. 701–710 (2014)
9. AustRoads guide to road safety part 3: speed limits and speed management. (316)
10. Jepsen, T., Jensen, C., Nielsen, T., Torp, K.: On network embedding for machine learning on road networks: a case study on the Danish road network. In: 2018 Proceedings - 2018 IEEE International Conference on Big Data, Big Data, pp. 3422–3431 (2019)
11. Yu, S., Rokka Chhetri, S., Canedo, A., Goyal, P., Faruque, M.: Pykg2vec: a python library for knowledge graph embedding. ArXiv Preprint ArXiv:1906.04239 (2019)

Enhancing Understandability of Omics Data with SHAP, Embedding Projections and Interactive Visualisations

Zhonglin Qu[1]([✉]) [iD], Yezihalem Tegegne[1], Simeon J. Simoff[2], Paul J. Kennedy[3], Daniel R. Catchpoole[4,5,6], and Quang Vinh Nguyen[2]

[1] School of Computer, Data and Mathematical Sciences, Western Sydney University, Penrith, Australia
{18885806,19201971}@student.westernsydney.edu.au

[2] MARCS Institute and School of Computer, Data and Mathematical Sciences, Western Sydney University, Sydney, Australia
{S.Simoff,q.nguyen}@westernsydney.edu.au

[3] Australian Artificial Intelligence Institute, Faculty of Engineering and Information Technology Sydney, Sydney, NSW, Australia
Paul.Kennedy@uts.edu.au

[4] The Tumour Bank, Children's Cancer Research Unit, Kids Research, The Children's Hospital at Westmead, Westmead, Australia
daniel.catchpoole@health.nsw.gov.au

[5] The Discipline of Paediatrics and Child Health, The Faculty of Medicine, The University of Sydney, Camperdown, Australia

[6] Faculty of Information Technology, The University of Technology Sydney, Sydney, Australia

Abstract. Uniform Manifold Approximation and Projection (UMAP) is a new and effective non-linear dimensionality reduction (DR) method recently applied in biomedical informatics analysis. UMAP's data transformation process is complicated and lacks transparency. Principal component analysis (PCA) is a conventional and essential DR method for analysing single-cell datasets. PCA projection is linear and easy to interpret. The UMAP is more scalable and accurate, but the complex algorithm makes it challenging to endorse the users' trust. Another challenge is that some single-cell data have too many dimensions, making the computational process inefficient and lacking accuracy. This paper uses linkable and interactive visualisations to understand UMAP results by comparing PCA results. An explainable machine learning model, SHapley Additive exPlanations (SHAP) run on Random Forest (RF), is used to optimise the input single-cell data to make UMAP and PCA processes more efficient. We demonstrate that this approach can be applied to high-dimensional omics data exploration to visually validate informative molecule markers and cell populations identified from the UMAP-reduced dimensionality space.

Keywords: Dimensionality reduction · Explainable AI · UMAP · Machine learning · PCA · Permutation importance · Random forest · SHAP · Visualisation

L. A. F. Park et al. (Eds.): AusDM 2022, CCIS 1741, pp. 58–72, 2022.
https://doi.org/10.1007/978-981-19-8746-5_5

1   Introduction

Vast amounts of Single-Cells RNA data have been collected and can be analysed based on the arrangement of populations of cells, such as cell type clusters. Single-cell sequence technologies can look at each of the individual cells. However, there is a limitation in the real-time live tracking [1]. Another challenge is determining subclone genotypes in more detail when integrating this with the spatial location of single cells. An essential task in analysing high-dimensional single-cell data is to find low-dimensional representations that capture the salient biological signals and render the data more interpretable and amenable to further analyses [2]. Current analysis methods rely on clustering or cell type assignment to identify the groups of cells [3]. Principal Component Analysis (PCA) [4] is one standard matrix factorisation method that can be applied to scRNA-seq data.

Dimensionality reduction (DR) is a common data pre-processing step that is usually necessary for classical machine learning analysis of high-dimensional data. DR transforms complex high-dimensional data into meaningful low-dimensional data with a feature transformation step that aims to minimise information loss before the analysis proceeds. The reduced dimension space must have correspondence to the intrinsic data dimensionality [5] for preserving both the accuracy and interpretability of the data analysis results. DR is also frequently used for data compression, exploration, and visualisation. Low-dimensional data representations remove noise but retain the signal of interest to understand hidden structures and patterns [6]. DR methods are classified as linear methods such as PCA, and non-linear methods such as t-distributed stochastic neighbour embedding (t-SNE) [7], and Uniform Manifold Approximation and Projection (UMAP) [8].

As a linear transformation method, PCA performs well when a dataset's underlying structure is known to be relatively linear. In contrast, UMAP, a relatively recent nonlinear embedding transformation method, is claimed to be more scalable and more accurate in identifying subpopulations in the cohorts than other methods for pattern visualisation and analysis [9]. For example, UMAP has been used to reveal special cells in single-cell transcriptomics, such as human T cells [10]. UMAP requires a higher computational cost than PCA, however, its performance can be improved significantly by using GPU acceleration [11]. As the nature of the non-linear method, it is hard to explain or interpret UMAP's outcomes and processes. The complexity of the statistical and mathematical UMAP model prevents domain users from understanding the clustering and projection results [12].

There are rising explainable machine learning methods focuses on solving outcome explanation problems, and explain the feature importance [13]. Local Interpretable Model-agnostic Explanations (LIME) and SHAP (SHapley Additive exPlanations) are two popular ones. LIME offers a method to faithfully explain any classifier or regressor's predictions with an interpretable model [14]. SHAP is a model-agnostic framework for interpreting predictions based on estimating the contribution of each feature for a particular prediction [15]. SHAP was created based on the principles from the cooperative game theory named coalitional games [16]. Shapley values were used in game theory to fairly distribute surplus across a coalition of players [17]. In the SHAP model, input features and Shapley values replace players to show the contribution to a given predicted result. In genomic studies, SHAP has been used to identify biomarkers [18];

for example, DeepSHAP [19] is a model used to explain the predictions of deep neural networks. SHAP is one of the algorithm transparency learning models to find the most relevant features contributing to the results [20]. Visual aids [21], such as interactive scatter plots and heat-maps, can also be used to allow users to view and interact with their inner functioning and improve their trustworthiness [20, 22].

Unlike the existing work on using PCA and scatter plots for explaining Convolutional Neural Network feature responses Field [23], we utilise PCA, and linkable interactive scatter plots to explain the UMAP method in our application. Our work was motivated by the fact that the PCA method has been used and understandable by many medical domain users. At the same time, the less interpretable UMAP [8] is better at effectively identifying similar subgroups in large high-dimensional datasets [9, 11]. By mapping the same cells to both PCA and UMAP and linking them together, we can find the original meaningful features in the single cell data with PCA and identify the similar cells with a better projection method UMAP. To better apply PCA and UMAP, we optimise the data with explainable machine learning SHAP on RF [23] for feature ranking and selecting the smaller number of dimensions. The SHAP model on RF can find the possible important markers based on their contribution to the RF results for further investigation of the diseases. Interactive and linkable scatter plots are created to find the sub-group of the cells and drill down to the raw data to improve the trustworthiness.

This paper contributes a novel method that uses a linear and interpretable projection method (such as PCA) to understand a more efficient but non-linear and un-interpretable projection method (such as UMAP) for better omics data analysis and uses the SHAP method to optimise the input data. Our specific contributions are:

i)   Contribute an interactive and multiple-view visualisation that links PCA's and UMAP's outputs for a better comparison and explanation,
ii) Use a machine learning explainable method, called SHAP, to optimise the raw data and find the most important markers of omics data with hundreds to thousands of dimensions. This process enables more effective projection results from DR methods which are not usually effective for the high number of dimensional data,
iii) Provide an innovative way to allow users to better understand the identified sub-populations in the omics data. The users can drill down to the raw data from the visualisations to explore the selected subpopulation of cells.

The paper is structured as follows. Section 2 describes our framework. The comparison can be made directly from the UMAP and PCA's results or via a feature selection process before the dimensionality reduction methods. Section 3 presents technical details of the SHAP, UMAP and PCA visualisations illustrated with two raw datasets. Section 4 shows optimisation of the input data to improve the outputs with SHAP run on random forest results. Section 5 validates the results from the above process and re-runs the UMAP and PCA with the optimised dataset. Section 6 provides the conclusion and future work.

2 Framework for Using SHAP to Optimise UMAP and PCA Input Data

Our framework is presented in Fig. 1, where we use a more interpretable and linear model, PCA, to understand a complex and non-linear model, UMAP. The framework includes three parts: A) interactive visualisations applied to all the dimensions for roughly comparing UMAP and PCA to find the similar patterns of both methods; B) as there are hundreds of dimensions in the datasets, the SHAP method run on Random Forest model is used to find the essential dimensions or features; and C) verify the SHAP results and reapplied PCA and UMAP visualisations with the results of B to get more accurate results. The interactive visualisations with the same subgroup cells highlighted in multiple views and the SHAP method with verification are used to ensure the results are correct and effective. Visualisations are developed with Python [24], Tableau Field [21], and Tabpy server from Tableau to support the explainable process.

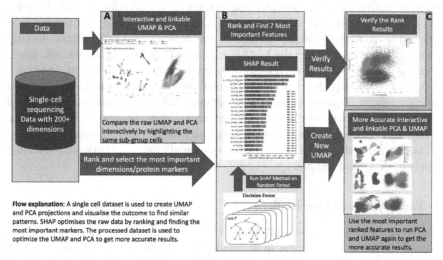

Fig. 1. The model for understanding a complex method UMAP with familiar PCA and SHAP method on Random Forest. A) interactive visualisations applied to all the 200+   dimensions for roughly comparing UMAP and PCA to find the similar patterns of both methods, B) SHAP method run on Random Forest model is used to find the most critical dimensions, and C) verify the SHAP results and reapply PCA and UMAP visualisations with the results of B to get more accurate results.

2.1 Initial Visualisations from the UMAP and PCA Projection Methods

Interactive visualisations and other supportive data analysis visualisations are created for preliminary comparing UMAP and PCA outcomes to find the visual similarities between the two methods. We run an experiment on a benchmark dataset, "NIST handprint Digitals" [25], to verify the similarities between the two projection methods, shown in

Sect. 3.3. Interactive visualisations based on UMAP and PCA results are developed to show visual analytics. Visualising the genomic data exploration process enables users to steer and drive the computational algorithms. User interactions with the system are designed and implemented as mechanisms by which users can augment the visualisation parameters, filter data, and other direct changes to the application.

Data visualisation tool Tableau [26] is used to develop interactive visualisations. Tableau provides the Tabpy server connection with Python and can directly use the Python analysis results for its interactive visualisations. These interactive visualisations can also help users to drill down their interesting subgroup's raw data, and to find the pattern similarities between PCA and UMAP and insights between the sub-group and the whole group.

2.2 Explainable Machine Learning SHAP for Important Feature Selection

The used single-cell total-seq data datasets have more than 200 dimensions. To improve the output quality, the explainable machine learning for feature relevance SHAP method runs on the Random Forest model. It is used to find the most important dimensions, such as the top seven highest-ranking features or dimensions in our experiments. Too high dimensional data could affect the UMAP accuracy. Therefore, an explainable feature relevance model SHAP is used on top of Random Forest results to rank all the features. The data have the sample level labels "Treatment" and "Titration", which can work as the prediction targets. Random forest uses the sample level labels as the prediction targets and randomly creates many classical decision trees to train the models. Shapley values are used as feature attributions and are a weighted average of all possible differences.

The purpose of this step is to find the most essential features and to use these features to apply to the UMAP method again to optimise the results.

2.3 Final Optimised Visualisations

The final visualisations can be used to review the SHAP results and re-apply interactive PCA and UMAP visualisations with the results of the SHAP ranking. After ranking the features to reduce the number of features from hundreds to much lower, we verify the effectiveness of the results by plotting the scatter plot with the most essential features, such as the first two highest rankings. The most important two features here are the biomarkers, which should be the ones that contribute to the target labels the most. Then we can validate the results by plotting the scatter plot with the two features and colouring them with target labels to see if the cells are separated better. If the results produce clearer boundaries in the scatter plot, then we conclude that the SHAP method to rank features is effective. As the domain experts might gain knowledge on the essential markers based on their domain knowledge, they can further evaluate and confirm the outcomes from the visualisation. After the verification, the first several essential features are used again iteratively in the PCA and UMAP process for further refinement. With these selected and fewer dimensions, the dimensional reduction methods can run faster. The results could be more accurate as we only use the most important features as the input data.

3 UMAP and PCA Visualisations

3.1 Datasets

We use a single-cell sequencing dataset generated by a total-seq platform [27] in our experiments to illustrate our proposed approach's utility and evaluate its performance. Total-seq, like CITE-seq (cellular indexing of transcriptomes and epitopes by sequencing [28]), is a single-cell sequencing technology that can measure both cell surface proteins (usually measured by cytometry) and transcriptomes at a single-cell level as by single-cell RNA sequencing (scRNA-seq). Therefore, the number of dimensions of a total-seq dataset can be higher than flow cytometry or RNA sequencing data.

The dataset has 12 deidentified human PBMC samples by combining four treatment/stimulation conditions and three titration levels. The four treatments include "Control" (control PBMC: no stimulation or cultured overnight), "IFNgTNFa" (IFN-gamma and TNF alpha stimulation), "IL2" (IL2 stimulation), and "ON" (Cultured Overnight). The three titration levels are "0.25x", "1x", and "4x". For each comparison, we merged the 12 samples by concatenation, resulting in a table with 24,410 rows and 202 columns. Each row corresponds to a single cell with quantitative measurements of the surface protein expressions (e.g., CD3, CD19 etc.), and each column name is a molecule marker. The empty values in the dataset are replaced with zero.

We also use a benchmark dataset "NIST handprint Digitals" [25]. The dataset contains images of handwritten digits with ten classes where each class refers to a number (0 to 9), 5620 instances, 64 attributes, and an 8×8 image of integer pixels in the range [0..16].

3.2 How Do PCA, UMAP, and SHAP Work?

PCA, UMAP and SHAP are used to analyse the same dataset from different angles. By comparing UMAP and PCA results, we can find similar projection and clustering patterns between both methods. PCA changes an $n * p$ original data matrix X to two matrices T (T ~ (n, p)) and P (P ~ (p, k)):

$$X = TP^T + E \tag{1}$$

E ~ (n, p) is the residual matrix, and n, p, and k are the number of samples, variables, and components.

$$\{T, P\} = argmax_{T,P}\left(\left\|X - TP^T\right\|_2^2\right) \tag{2}$$

The combination of T and P vectors is referred to as principal components that capture as much as the variance in the original data in the least squares sense. UMAP use the closest fuzzy topological structure to find the low dimensional representations. Let E as the set of all possible 1-simplices, and $w_h(e)$ is the weight of the 1-simplex e in the high dimensional case, and $w_l(e)$ is the weight of e in the low dimensional case, then the cross entropy will be

$$\sum_{e \in E} w_h(e) log(\frac{w_h(e)}{w_l(e)}) + (1 - w_h(e)) log(\frac{1 - w_h(e)}{1 - w_l(e)}) \tag{3}$$

This force-directed graph layout algorithm minimises the cross entropy and lets the low dimensional representation settle into a state that accurately represents the original multidimensional data. SHAP is used on the Random Forest to get the essential features. The SHAP method requires retraining the model on all feature subsets. Let $x_i \in \mathbb{R}^p$ denotes an input datapoint and $f(x_i) \in \mathbb{R}$ denotes the corresponding output of function f. Shapley values decompose this number into a sum of feature attributions: $f(xi) = \sum_{j=0}^{p} \varnothing_j$. $\varnothing_0$ denotes a baseline expectation and $\varnothing_j$ ($j \geq 1$) denotes the weight assigned to feature x_j at point x_i. Let $v: 2^p \rightarrow \mathbb{R}$ be a value function such that $v(S)$ is the payoff associated with feature subset $S \subseteq [p]$ and $v(\{\varnothing\}) = 0$. The Shapley value ϕ_j is given by j's average marginal contribution to all subsets that exclude it:

$$\phi_j = \frac{1}{p!} \sum_{s \subseteq [p] \setminus \{j\}} |S|!(p - |S| - 1)! \big[v(S \cup \{j\}) - v(S) \big] \tag{4}$$

Shapley values are the average of all the marginal contributions to all possible coalitions, see full details in [17]. Here Shapley values are used as feature attributions and are weighted the average of all possible differences. These important features are verified with scatter plot visualisations and domain experts' knowledge. Then the essential features are used to rerun UMAP and PCA to improve the efficiency, accuracy, and comparisons.

3.3 Similar Projection and Clustering Patterns Between PCA and UMAP

We used the above datasets "NIST handprint Digitals" and the single cell dataset scRNA-seq to define why it is possible to use PCA to explain UMAP. PCA and UMAP are used to reduce the dimensions and plot the two projected principal components (PCs). We visualise the PCs interactively as scatter plots in Tableau software[1].

As you can see, PCA is good for preserving the linear positions while UMAP can cluster similar marks better. Figure 2 shows the result of both PCA and UMAP methods for the "NIST handprint Digitals" dataset. By highlighting the same digit number in both PCA and UMAP (see Fig. 2), we found that the same number had similar patterns in both PCA and UMAP. For example, Fig. 2 shows that number "1" has two groups of projections in PCA and similarly in UMAP (Fig. 2-A); number "9" projects sparsely in PCA, and it also clusters in a different group in UMAP (Fig. 2-C), and number "0" and "6" project together in PCA and cluster well in UMAP as well (Fig. 2-B and Fig. 2-D).

For the single-cell RNA sequencing (scRNA-seq) data, the UMAP method reduces 202 markers to two principal components (PCs) and plots as scatter plots visualisations. The labels "Treatment" and "Titration" are combined as one label and are mapped to 12 colours. We then compare UMAP and PCA in an interactive multiple-view dashboard by plotting the data file with both UMAP and PCA algorithms and colouring them with the 12 combinations of Treatment and Titration. If we highlight two subsets: IL2,0.5x, and Control,0.25x, we can find they are in similar projection and clustering patterns, as shown in Fig. 3.

---

[1] https://www.tableau.com/.

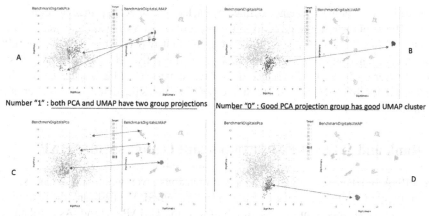

Fig. 2. A dashboard on PCA (left image) and UMAP (right image) for NIST handprint digitals benchmark dataset shows similar projection and cluster patterns where A, B, C and D highlights selected digits "1", "0", "9" and "6" in the charts respectively.

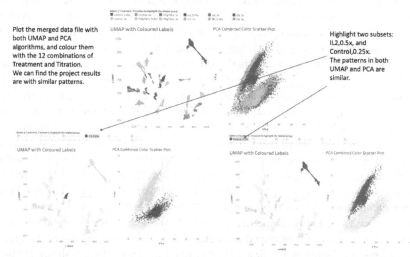

Fig. 3. Compare UMAP and PCA in an interactive multiple-view dashboard. The clustering patterns of the highlighted subgroups of cells in UMAP and PCA are similar. If we highlight a subset of cells, for example, IL2,0.5x, and Control,0.25x, we can find they are in similar projection and clustering patterns.

The above comparisons show that PCA and UMAP results have similar clustering and projection patterns. We then hypothesise that using domain users' familiar algorithm PCA may be possible to understand the new and more complex but more accurate UMAP results. In that case, we can then improve the use of UMAP in single-cell data visualisations and interpretations. PCA and UMAP components and their significance can be explained using "explained variance" and "explained variance ratio". Explained variance is the amount of variance defined by each of the selected principal components (PCs). The explained variance ratio is the percentage of variance explained by each selected feature. If we only keep three PCs, the first two PCs of UMAP cover 69% of all the variance while PCA covers 82%. If we keep five PCs, the first two PCs of UMAP cover 48% of all the variance while PCA covers 68%. PCA keeps more values of all attributes than UMAP, and keeping fewer PCs makes the first two PCs cover more variances.

4   Rank and Select the Most Important Features with SHAP

The scRNA-seq data have a high number of dimensions (over two-hundred dimensions), which might affect the accuracy of the projection method. We first rank the features to identify the most important ones to run the following UMAP method. As the data has labels Treatment and Titration, we can also run a supervised machine learning model, Random Forest. Explainable machine learning model SHAP is used on top of the Random Forest to find the most important markers. We then plot the most important twenty markers as bar charts and rank the features. The domain experts' knowledge and a scatter plot for the first two markers are used to confirm the results. Then UMAP and PCA are run again based on the first several most essential markers to get a more accurate visualisation. Interactive scatter plot visualisations with brushing, and linked techniques [29] are developed for users' drilling down their exciting features.

In our experiment, we randomly built 100 decision trees and then ran SHAP (see more details in Sect. 3.2) to calculate the average of all the values predicted by all the trees in the forest. The new record is assigned to the category that wins the majority vote. As we have 12 samples, first, we combined four Treatment and three Titration and let the 12 combinations be the labels for each cell. Then we can train a random forest model with 100 decision trees. SHAP method is used to rank the 202 attributes or markers as shown in Fig. 4. In the results, we show the first 20 important attributes, and we want to choose the first seven of them to apply dimensionality reduction again because more than seven attributes are very slow to run on Tabpy server between Python and Tableau. From our experiments, the seven most important features can keep the result accurately, and the performance is well at the same time. In the SHAP method, the permutation-based importance can be used to overcome drawbacks of default feature importance computed with mean impurity decrease. The first several most important markers as shown in Fig. 4 also are plotted with more details, showing the feature value.

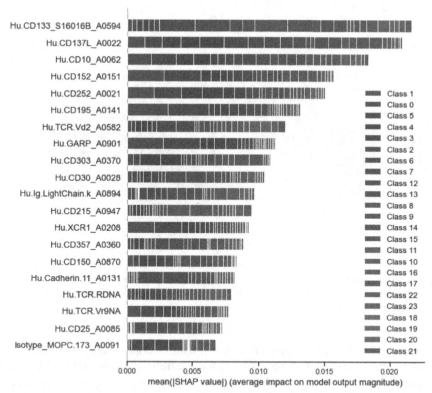

Fig. 4. Use explainable machine learning model SHAP on top of the Random Forest result and labels as target to rank all the markers and show the first important 20 markers.

5   Validation of the SHAP Results

To verify the outcomes of the SHAP results, we used the results to plot the first two important features with scatter plot visualisations and the domain expert's confirmation. Another purpose of this research is to find the most important biomarkers that can be affected by the Treatment methods and Titration. The SHAP and the random forest can rank all the biomarkers based on the sample level label Treatment methods and Titration. If the ranking is right, we could get the most important biomarkers.

We used two ways to validate the ranking results. One way is to use domain users' knowledge as they already know which biomarkers are the most important ones. Another way is to plot the first two important biomarkers as a scatter plot and coloured the markers based on the Treatment label and the Titration label. If the two biomarkers are the most important ones that affect the label Treatment and Titration, then the boundaries among these groups should be more obvious than only randomly choosing two biomarkers as we use the targeted labels to colour the cells.

We asked one domain expert to verify the results, and the expert confirmed that the first two features are the ones that have been known as the biomarkers that affect the Treatment and Titration labels. We then plot the scatter plot and get the results are

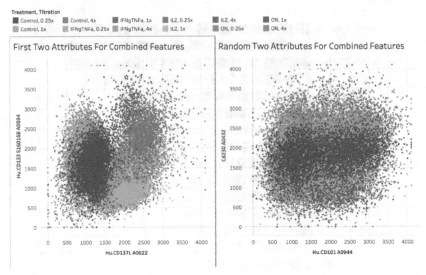

Fig. 5. Plot the first two important markers after ranking all the markers with SHAP (Left plot). The scatter plot with the selected attribute has clearer boundaries than the one from the random chosen two attributes (Right plot).

shown in Fig. 5 (left), and another scatter plot with the random chosen two attributes and coloured with 12 combined features. We can see that the scatter plot on the left has clear boundaries based on the target colours. It proves that the first two important markers can make the groups be separated well with clear borders. This could potentially help the analysts to identify the useful subpopulations of the cells for further analysis. With the two verification methods, we can conclude that the explainable machine learning model SHAP can find the actual most important markers.

We then use the seven most important markers to apply PCA and UMAP to create interactive plots. The fewer dimensions can make the calculation and visualising process more efficient than the original 202 dimensions. The results are shown in Fig. 6. In Fig. 6 top, we can see that the UMAP and PCA results all have better clustering and projection based on the target colours than the original results in Fig. 3. The plotted visualisations are interactive, and we can drill down each group of samples to the raw data. For example, if we select Treatment as "Control" and Titration as "0.25" (see Fig. 6 bottom), then both scatter plots will highlight the same group of cells. On our interactive dashboard, we can directly show the raw data of this subgroup of the cells as shown in Fig. 7. We select a group of cells and can select "View Data", then we can find the raw data of these selected cells in a pop-up window.

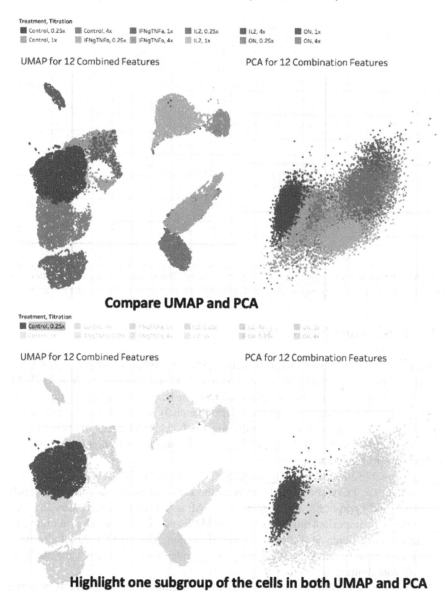

Fig. 6. The interactive and linkable dashboard after applying the most important markers to the UMAP and PCA. The top shows the projection coloured with different labels, while the bottom shows one highlighted subgroup of cells. (Color figure online)

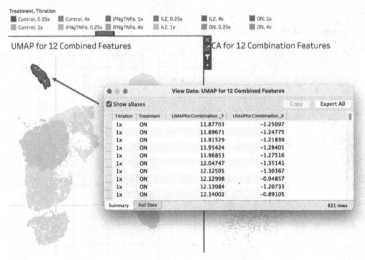

Fig. 7. Select a Group of Cells and Show the Raw Data. The linkable multi-view scatter plot allows the user to choose a group of clustered cells and choose to check the raw data in a table.

6   Conclusion and Future Work

This paper uses a domain user's familiar method, PCA, to compare and understand a new and complex method, UMAP. We created interactively and linkable PCA and UMAP scatter plots. The linking and brushing interactions can help users drill down the subsets of each sample and directly go to the raw data. The analysis results and the interactive comparison can help domain users trust the new method UMAP based on their familiar method PCA results. The results can also help the domain users further investigate and assist the decision-making.

An explainable machine learning model SHAP on Random Forest is used to optimise the input data for PCA and UMAP because the dataset has the sample level labels "Treatment" and "Titration". The purpose of SHAP is to rank all the markers and then only use the most important attributes to plot UMAP and PCA again. This can improve the accuracy and efficiency of dimensionality reduction. The SHAP method also finds the most important biomarkers affected by the sample labels, which is very useful in single-cell data domain analysis.

In the future, we will i) use other explainable machine learning methods on more datasets to find important attributes for multidimensional data, for example, GINI, LIME; ii) we will make more comparisons among the different dimensionality reduction methods and find meaningful features to help elucidate novel biomarkers in the continuum of cells for the domain users to understand the new machine learning models and trust the models eventually; iii) we will also find new ways to combine different machine learning models, including supervised and unsupervised models, to analyse big and complex genomic data.

Acknowledgement. We appreciate Yu "Max" Qian of J. Craig Venter Institute and the BioLgend Company for providing the TotalSeq dataset used in the paper. All datasets used in the study are fully de-identified and do not contain any protected health information.

References

1. Wong, K.-C.: Big data challenges in genome informatics. Biophys. Rev. **11**, 51–54 (2018)
2. Pierson, E., Yau, C.: ZIFA: dimensionality reduction for zero-inflated single-cell gene expression analysis. Genome Biol. **16**(1), 241 (2015)
3. Yang, Y., et al.: SAFE-clustering: single-cell Aggregated (from Ensemble) clustering for single-cell RNA-seq data. Bioinformatics **35**(8), 1269–1277 (2019)
4. Hosoya, H., Hyvärinen, A.: Learning visual spatial pooling by strong PCA dimension reduction. Neural Comput. **28**(7), 1249 (2016)
5. Sumithra, V.S., Subu, S.: A review of various linear and non linear dimensionality reduction techniques. Int. J. Comput. Sci. Inf. Technol. **6**(3), 2354–2360 (2015)
6. Nguyen, L.H., Holmes, S.: Ten quick tips for effective dimensionality reduction. PLoS Comput. Biol. **15**(6), e1006907 (2019)
7. Konstorum, A., et al.: Comparative analysis of linear and nonlinear dimension reduction techniques on mass cytometry data. bioRxiv, p. 273862 (2018)
8. Etienne, B., et al.: Dimensionality reduction for visualizing single-cell data using UMAP. Nat. Biotechnol. **37**(1), 38–44 (2018)
9. Trozzi, F., Wang, X., Tao, P.: UMAP as a dimensionality reduction tool for molecular dynamics simulations of biomacromolecules: a comparison study. J. Phys. Chem. B **125**(19), 5022–5034 (2021)
10. Szabo, P.A., et al.: Single-cell transcriptomics of human T cells reveals tissue and activation signatures in health and disease. Nat .Commun. **10**(1), 4706–4716 (2019)
11. Tegegne, Y., Qu, Z., Qian, Y., Nguyen, Q.V.: Parallel nonlinear dimensionality reduction using GPU Acceleration. In: Xu, Y., et al. (eds.) AusDM 2021. CCIS, vol. 1504, pp. 3–15. Springer, Singapore (2021). https://doi.org/10.1007/978-981-16-8531-6_1
12. Wang, Y., et al.: Understanding how dimension reduction tools work: an empirical approach to deciphering t-SNE, UMAP, TriMAP, and PaCMAP for data visualization (2020)
13. Nauta, M., et al.: From anecdotal evidence to quantitative evaluation methods: a systematic review on evaluating explainable AI. arXiv preprint arXiv:2201.08164 (2022)
14. Ribeiro, M.T., Singh, S., Guestrin, C.: "Why should i trust you?": explaining the predictions of any classifier. In: International Conference on Knowledge Discovery and Data Mining, pp. 1135–1144. ACM (2016)
15. Lundberg, S.M., Lee, S.-I.: A unified approach to interpreting model predictions. In: Proceedings of the 31st International Conference on Neural Information Processing Systems, pp. 4768–4777. Curran Associates Inc., Long Beach (2017)
16. Osborne, M.J.: A Course in Game Theory. In: Rubinstein, A. (ed.) MIT Press, Cambridge (2006)
17. Shapley, L.S., Kuhn, H., Tucker, A.: Contributions to the theory of games. Ann. Math. Stud. **28**(2), 307–317 (1953)
18. Watson, D.: Interpretable machine learning for genomics (2021)
19. Fernando, Z.T., Singh, J., Anand, A.: A study on the interpretability of neural retrieval models using DeepSHAP. In: Proceedings of the 42nd International ACM SIGIR Conference on Research and Development in Information Retrieval (2019)
20. Vilone, G., Longo, L.: Explainable artificial intelligence: a systematic review. arXiv preprint arXiv:2006.00093 (2020)

21. Strobelt, H., et al.: Lstmvis: a tool for visual analysis of hidden state dynamics in recurrent neural networks. IEEE Trans. Visual Comput. Graphics **24**(1), 667–676 (2017)
22. Thelisson, E.: Towards trust, transparency and liability in AI/AS systems. In: IJCAI (2017)
23. Dimitriadis, S., Liparas, D.: How random is the random forest? Random forest algorithm on the service of structural imaging biomarkers for Alzheimer's disease: from Alzheimer's disease neuroimaging initiative (ADNI) database. Neural Regen. Res. **13**(6), 962–970 (2018)
24. Python (2020). https://www.python.org/
25. Candela, M.G.J.B.G., et al.: NIST form-based handprint recognition system. Technical Report NISTIR 5469, Nat'l Inst. of Standards and Technology 91994)
26. Tableau (2020). https://www.tableau.com/
27. BioLegend: Comprehensive solutions for single-cell and bulk multiomics (2021). https://www.biolegend.com/en-us/totalseq?gclid=CjwKCAjwx8iIBhBwEiwA2quaq0V-IkCRsY9UZ6G1Lop5Tfd0dl1m_YF-_fyd-1Hgz5fUvpEvevRpcRoCIjUQAvD_BwE. Accessed 22 Aug 2021
28. Stoeckius, M., et al.: Simultaneous epitope and transcriptome measurement in single cells. Nat. Methods **14**(9), 865–868 (2017)
29. Radoš, S., et al.: Towards quantitative visual analytics with structured brushing and linked statistics. Comput. Graph. Forum **35**(3), 251–260 (2016)

WinDrift: Early Detection of Concept Drift Using Corresponding and Hierarchical Time Windows

Naureen Naqvi[✉], Sabih Ur Rehman, and Md Zahidul Islam

Charles Sturt University, Bathurst, NSW, Australia
{nnaqvi,sarehman,zislam}@csu.edu.au

Abstract. In today's interconnected society, large volumes of time-series data are usually collected from real-time applications. This data is generally used for data-driven decision-making. With time, changes may emerge in the statistical characteristics of this data - this is also known as *concept drift*. A concept drift can be detected using a concept drift detector. An ideal detector should detect drift accurately and efficiently. However, these properties may not be easy to achieve. To address this gap, a novel drift detection method *WinDrift (WD)* is presented in this research. The foundation of WD is the early detection of concept drift using corresponding and hierarchical time windows. To assess drift, the proposed method uses two sample hypothesis tests with Kolmogorov-Smirnov (KS) statistical distance. These tests are carried out on sliding windows configured on multiple hierarchical levels that assess drift by comparing statistical distance between two windows of corresponding time period on each level. To evaluate the efficacy of WD, 4 real datasets and 10 reproducible synthetic datasets are used. A comparison with 5 existing state-of-the-art drift detection methods demonstrates that WinDrift detects drift efficiently with minimal false alarms and has efficient computational resource usage. The synthetic datasets and the WD code designed for this work have been made publicly available at https://github.com/naureenaqvi/windrift.

Keywords: Concept drift detection · Time-series · Multiple window-based hierarchies · Multiple hypothesis tests · Kolmogorov-Smirnov statistical test

1 Introduction

In today's hyper-connected world, brought on due to the emergence of contemporary ways in which the world is operating such as smart cities, intelligent power grids, smart farms for precision agriculture, efficient water conservation, and smart waste management etc.; large volumes of time-series data are being produced at an astronomical rate [20]. The time-series data consists of a set of observations collected over time. The real-world applications generally use this data to learn in an incremental fashion using machine learning models. This process is also referred to as incremental learning. However, with time, the changes in the statistical characteristics of data can emerge between two time periods - this is also known as *Concept Drift* [11,13,33]. In incremental learning, the efficient and accurate detection of concept drift is an important task. In this way,

L. A. F. Park et al. (Eds.): AusDM 2022, CCIS 1741, pp. 73–89, 2022.
https://doi.org/10.1007/978-981-19-8746-5_6

Table 1. Symbols and notations.

Notation	Description
x_n	Input data arriving in streaming setting with n total number of observations for an attribute X
W	Window containing sample distribution where, W_H for historical, W_N for new data
μ	The mean used to randomly generate a sample distribution
σ	The standard deviation used to randomly generate a sample distribution
$F(x)$	Empirical Cumulative Distribution Function (ECDF) of a sample distribution
n	Total no. of samples in a window, where, n_H for historical, n_N for new data window
x	x^{th} observation in the distribution that is used to draw out a sample and calculate $F(x)$
T	A population distribution for X whose sufficient statistics will be used as a reference to detect drift
$W_{[H,N]}$	Sample distribution drawn from a population distribution T between time interval $[H, N]$
H_0	Null hypothesis which represents that there is no effect in the population
H_A	Alternative Hypothesis which represents that there is an effect in the population
α	Distance that the sample mean must be from the null hypothesis H_0
D_s	The statistical geometrical maximum vertical distance between the two sample distributions
D_c	The critical value to test if the two sample distributions have dissimilarities between them
$C_{\{1,..,14\}}$	The data patterns used in the synthetic datasets i.e., $C_{\{1,..,10\}}$ and real datasets i.e., $C_{\{11,..,14\}}$

the detector can inform the incremental learning model about the change in the statistical characteristics of the underlying time series data so that the model can be updated based on emerging concepts. If this drift is not efficiently and accurately detected, it may degrade the performance of a model. The performance degradation could be due to the missed opportunity to update the model despite the emergence of drift (due to a *False Negative* outcome from the drift detector) or to incorrectly update the model when no drift exists (due to a *False Positive* outcome from the drift detector) [12, 24].

Several approaches to concept drift detection in non-time series data that exist in the literature are discussed in this section. A state-of-the-art approach to drift detection uses windows for comparison. These detectors divide the incoming data into two or more windows based on some criteria and compare their statistics to detect drift [4, 5, 10–12, 15]. In another approach, statistical analysis is used for drift detection; these methods use mathematical inequalities or statistical distances to analyze the dissimilarities between the two distributions [8, 11, 14–16, 26]. Another significant approach uses statistical hypothesis tests. In this approach, the detector uses statistics based hypothesis tests to detect drift [1–3, 11, 25, 30, 34]. Although the above-mentioned methods have proven useful to drift detection, they present certain performance challenges, especially for time-series data [18] such as the *delay in detection*, reporting a *high number of false alarms* and requiring high *computational resources*.

Motivated by the need to build an efficient and accurate drift detector for time-series data, we propose a novel method called *WinDrift*. Our method uses multiple hierarchical and corresponding windows to assess drift into two modes i.e., *Mode I: Consecutive Mode* (to detect drift between the consecutive windows), and *Mode II: Corresponding Mode* (to detect drift between corresponding windows). The window sizes are chosen by taking user input on the granularity of time-windows. For drift assessment, the statistical characteristics of the sample distributions are compared to determine their dissimilarity which represents *drift*. Our key contributions are summarized as follows:

- Designed a novel method, for time-series data to detect drift into two modes i.e., consecutive mode (between adjacent windows) and corresponding mode (between two windows of the corresponding time period that are a cycle apart).
- Designed to achieve early drift detection by using multiple hierarchical sliding windows to assess drift between corresponding time periods.
- Evaluated the performance of WinDrift against ten diligently designed synthetic datasets and four real datasets to demonstrate that WinDrift detects drift efficiently with two key objectives: a *high detection performance* i.e., a high number of true detection with a low number of false alarms (*Objective # 1*) and using the *least computational resources* i.e., CPU and memory usage (*Objective # 2*).
- Released the code and synthetic datasets at https://github.com/naureenaqvi/windrift.

The remainder of the article is organized as follows: Sect. 2 provides preliminary information about the fundamental concepts used in this work. Section 3 presents the detailed proposed method with the help of an algorithm and a step-by-step guide. Experimentation and results are presented in Sect. 4. Section 5 concludes this work.

2   Preliminaries

In this section, background information related to the implementation of WinDrift is presented. Readers are referred to Table 1 which presents the symbols and notations used throughout this article. In the literature [32], a concept drift is characterized as a change that emerges in a data distribution between two time periods. A data distribution of a population can usually contain one or more sample distributions. In a statistical significance test, we may conduct a two sample hypothesis test using one or more statistical characteristics of a sample distribution. Here, the intention is to test sample distributions based on a certain characteristic and confirm if they belong to identical population distribution. This forms the basis of the test we apply to detect *drift*.

In this article, we measure statistical distances between the two sample distributions [14] to determine if the samples, say W_H, from the historical window, and W_N, from the new data window are drawn from an identical population distribution T as shown in Fig. 1a. Using this analogy, a two-sample hypothesis test can be deemed suitable for drift detection. To measure the statistical distance between W_H and W_N, the two-sample Kolmogorov-Smirnov (KS) test [22,27,31] is used in this article. Here, the difference between the Empirical Cumulative Distribution Function (ECDFs) of the two sample distributions of a single attribute X is used to determine dissimilarities between them. To calculate the ECDF i.e., $\hat{F}_n(x)$ of a sample distribution as the observation x arrives, the following expression can be used [28]:

$$\hat{F}_n(x) = \frac{no.\ of\ observations\ in\ a\ sample\ distribution \leq x}{n}, \qquad (1)$$

where n is the number of observations in a sample distribution. We refer to this technique as *CalculateECDF* in this article. A graphical representation of the ECDFs generated by W_H i.e., $\hat{F}_H(x)$ and W_N i.e., $\hat{F}_N(x)$, and D_s is presented in Fig. 1d using blue, green, and red colour lines respectively. A null hypothesis H_0 for a KS test can be mathematically expressed as $D_s > D_c$, where, D_s represents the maximum statistical distance between the ECDFs of the two distributions given by:

$$D_s = \max_{x \in \mathbb{R}} |\hat{F}_H(x) - \hat{F}_N(x)|, \tag{2}$$

D_c is the critical value used as the maximum allowable threshold to test H_0. In KS, it is dependent on two factors i.e., the level of significance α, and the number of observations in each window W_H i.e., $\hat{n}_H(x)$ and W_N i.e., $\hat{n}_N(x)$ and it is given by:

$$D_c = \left(\sqrt{-\frac{1}{2} \left(\ln \frac{\alpha}{2} \right)} \right) \left(\sqrt{\frac{1}{n_H} + \frac{1}{n_N}} \right), \tag{3}$$

The test hypothesis described above can be summarized as follows:

$$\begin{cases} H_0 : drift\ not\ detected & if\ \ D_s < D_c \ \ then\ \ \{W_H, W_N\} \in T \\ H_A : drift\ detected & if\ \ D_s > D_c \ \ then\ \ \{W_H, W_N\} \notin T \end{cases} \tag{4}$$

The mathematical expressions presented in Eqs. 2, 3, and 4 form the basis of a technique called *AssessDrift*. As illustrated in Fig. 1d, we use this to assess the drift between two windows. In the subsequent section, we explain our proposed method.

3   The WinDrift (WD) Method

In this section, we start by presenting the key components of our method in Sect. 3.1. Furthermore, a working example of our method using a diligently designed synthetic dataset are presented in Sect. 3.2. Readers are encouraged to refer to Fig. 1 in conjunction with the discussions relating to the discussion on key components below.

3.1   Key Components

The key components of our proposed method are as follows:

  (i) setting up multiple time-based hierarchical levels
 (ii) selection of window sizes at each hierarchical level
(iii) categorization of assessment operation into two modes
 (iv) assessment of drift using a two-sample hypothesis test

  (i) **Setting up multiple time-based hierarchical levels** – WinDrift is designed to assess drift for time-series data using multiple time-based hierarchical levels. Acknowledging the differences between various applications, we seek user input to create sliding windows on each level. This input helps determine the following parameters which are necessary to perform steps required for achieving Component (i) above.
   – *number of levels (winLevSize)* in a time-hierarchy that use the window sizes on each level in order to define how frequently the drift assessment should occur;

- *maximum unit size (maxCycLen)* in a time-hierarchy that is used to define the maximum cycle length to start drift assessment between corresponding time periods;
- *minimum unit size (dataBlock)* contains the minimum block size of observations extracted from the arrival of a data stream; used to parse windows on each level.

(ii) **Selection of window sizes at each hierarchical level** – WinDrift takes both mandatory and optional inputs from the users. Out of which *winLevSize* is a mandatory input that creates hierarchical windows of various sizes on multiple levels. Other inputs are explained in Step 2 of Sect. 3.2. This input contains information about minimum level window i.e., *dataBlock*, intermediate windows, and maximum level window i.e., *maxCycLen*. In a *winLevSize* input, a *dataBlock* is always a single minimum unit of drift assessment. Generally, the users can have as many window levels as they require for as suitable for the data received from their applications. In the example presented in Fig. 1a, we arbitrarily select four levels of drift assessment, i.e., *winLevSize* in an array *{yearly, half-yearly, quarterly, and monthly}* as given by the user e.g., $\{12, 6, 3, 1\}$. The first element of the *winLevSize* demonstrates the *maxCycLen* i.e., $\{12\}$ in this instance; while each of the other elements represents the number of *dataBlocks* on each level.

This input follows a structure such that a higher-level window contains a nested level windows underneath. In Fig. 1a, Level 1 represents the lowest drift assessment level that creates windows on a monthly-basis. Level 2 assesses drift on a quarterly-basis; Level 3 assesses drift on a half-yearly basis. Finally, Level 4 assess drift on a yearly-basis. Note that the above-given example follows a specific user input; however, this may vary from application to application. For example, if a time-series is an average daily temperature of a location, then the four levels defined above may be suitable as it compares between corresponding months of two years e.g., Jan 2019 with Jan 2020; Feb 2019 with Feb 2020 and so on, to assess drift between Y2019 and Y2020. However, if a time-series is the average Higher Secondary School Certificate (HSC) score for a class of students taking Mathematics, then a suitable input may differ from the one defined above. Hence, user input is of essence in our method as it provides the flexibility to tune the detection frequency by acknowledging and accommodating for the user needs.

(iii) **Categorization of assessment operation into two modes** – In WinDrift, the two windows sizes *maxCycLen* and *dataBlock* are of strategic importance. The role of *maxCycLen* is to divide the drift assessment operations into two modes, i.e., *Mode I: Consecutive Mode*, and *Mode II: Corresponding Mode*. In *Mode I*, the drift is assessed between two consecutive windows on the minimum window level and all other intermediate window levels until *maxCycLen* is reached. In *Mode II*, the drift is assessed on all levels between two corresponding windows that exist a cycle distance apart. Although it is envisaged that most real-world applications should assess drift between two corresponding time periods, it is not always possible to have sufficient historical data available for windows comparison. For example, the big data collected from smart cities [18, 20] do not always have any reference or historical data when the applications first start operating. Therefore, *Mode I* is also introduced in our method. Since our method allows multiple hierarchical level so

---

Algorithm 1: WinDrift: Corresponding and hierarchical time windows.

---

1 **Input:** Mandatory: *winLevSize[ ]*, and *datablocks* ; Optional: α, and *userContinue* – allows the user to continue
   to assess drift.
   Result: *driftFlag[ ]*.

2 **Method:**
3 **begin**
   winlevsize[ ] ← ∅;   ▷ the user-defined window sizes for drift assessment;
   dataBlock ← 0;   ▷ the batches of streaming data used for drift assessment
   dBcount ← 0;   ▷ the count of the dataBlocks received at any point in time
   driftFlag[ ] ← ∅;   ▷ the outcome of drift assessment across two modes
4 **end**

5 **Step 1: Get UserInput**
6 **begin**
7   **get** userInput {winlevsize[ ], α, userContinue};
8   **function** CountArray(r, winlevsize[ ])   ▷ return r ← count of the number of window levels
9   **function** SortArray(p, winlevsize[ ])   ▷ return p ← position of each of the window levels
10   **function** SearchArray(o, sortedWin[ ])   ▷ return o ← size of the maximum cycle length
11 **end**

12 **Step 2: Prepare Windows**
13 **begin**
14   *Divide sliding windows into* W_H *and* W_N *on each level.*
     i ← 0;   ▷ the unit size of a window extracted from the winlevsize array
     j ← 0;   ▷ the sliding window index is maintained separately for each level
     m ← 0;   ▷ the count of cycles after the maximum cycle length is reached
     k_1 ← i * j;   ▷ dataBlocks in W_{H1} for Mode I using the historical data
     l_1 ← i * (j + 1);   ▷ dataBlocks in W_{N1} for Mode I using new data
     k_2 ← ((m − 1) * o) + (k_1); ▷ dataBlocks in W_{H2} for Mode II using the historical data
     l_2 ← (m * o) + (k_1);   ▷ dataBlocks in W_{N2} for Mode II using new data
15 **end**

16 **Step 3: Detect Drift**
17 **while** *(userContinue = True)* **do**
     get dataBlock;   ▷ the arrival of data stream with a timestamp
     dBcount ← dBcount + 1;   ▷ increment of count of dataBlock by 1
     Check dBcount – recursively to determine the modes and window levels.
     if *dBcount < o* **then**

         LevelCheck uses the following three functions.
         SlidingWindow – slides the windows on each level by checking index.
         AssessDrift – is a two-sample KS test to assess drift using ECDF.
         CalculateECDF – calculates the ECDF of a data distribution

         Step 3a: Assess Drift (Consecutive Mode)
         Assess drift on all levels until the maximum cycle length is reached.
         for p ← 1 **to** < r **do**
             function LevelCheck(k_1,l_1);   ▷ uses k_1 and l_1 dataBlocks to slide windows and assess drift
             return driftFlag[ ];   ▷ drift in *Mode I* consecutive windows
             p ← p + 1;   ▷ increment until max-1 window size is reached
         end
     else
     end

     Step 3b: Assess Drift (Corresponding Mode)
     Assess drift & increment cycle number – once maxCycLen is reached.
     while n ≥ 1 **do**
         for p ← 1 **to** r **do**
             function LevelCheck(k_2,l_2);   ▷ uses k_2 and l_2 corresponding dataBlocks to assess drift
             return driftFlag[ ];   ▷ drift in *Mode II* corresponding windows
             p ← p + 1;   ▷ increment until max window size is reached
         end
         n ← n + 1;   ▷ increment as soon as maxcyclen is reached
     end
18 **end**

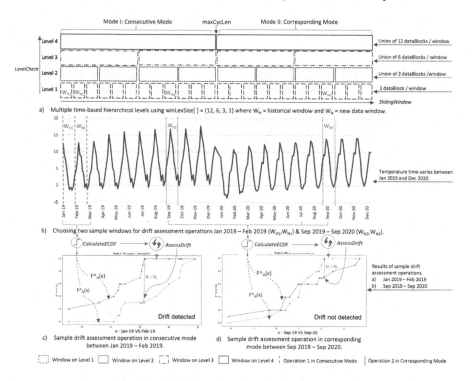

a) Multiple time-based hierarchical levels using winLevSize[ ] = {12, 6, 3, 1} where W_H = historical window and W_N = new data window.

b) Choosing two sample windows for drift assessment operations Jan 2019 – Feb 2019 (W_{H1},W_{N1}) & Sep 2019 – Sep 2020 (W_{H2},W_{N2}).

c) Sample drift assessment operation in consecutive mode between Jan 2019 – Feb 2019.

d) Sample drift assessment operation in corresponding mode between Sep 2019 – Sep 2020.

Fig. 1. WinDrift (WD) - Multiple Hierarchical Window based Drift Assessment Operations.

it allows the user to detect drift on the most granular level in *Mode II* as required by their applications. This allows drift detection between adjacent windows without waiting for the entire dataset to start assessment. All of these characteristics of WinDrift enable the *early detection* of drift. Note that these modes are mutually exclusive and do not occur simultaneously. To ensure symmetry in data parsing, it is advised that the size of *maxCycLen* should always be divisible by all other window sizes underneath. To assess drift, different window sizes are created on each level using one or union of *dataBlocks* as given in the *winLevSize*.

(iv) **Assessment of drift using a two-sample hypothesis test** – In WinDrift, we use a technique called *LevelCheck* that checks for a single *dataBlock* on the lowest level; while, a union of multiple *dataBlocks* is required on the higher levels. On each level, we create two windows, i.e., W_H and W_N. Both windows have the same size if they fall on the same level; however, they can take different sizes for different levels. We refer to this technique as *SlidingWindow*. In Fig. 1b we present time-series data to demonstrate WinDrift operations. Figure 1c and 1d show a graphical representation of the drift assessment process for *Mode I* i.e., Jan 2019–Feb 2019 (W_{H1},W_{N1}) and *Mode II* i.e., Sep 2019–Sep 2020 (W_{H2},W_{N2}) respectively. Next the window statistics are prepared using Eq. 1; and the drift is assessed using Eqs. 2 and 3, and 4. In the next sub-section, we present our working example using a synthetic dataset. This will aid in reinforcing the understanding of the key principles

Table 2. Sample data for Y2019.

2019											
H1						H2					
Q1			Q2			Q3			Q4		
J	F	M	A	M	J	J	A	S	O	N	D
12.50	9.10	7.20	4.50	0.70	−1.30	−0.60	−1.60	−0.30	2.60	4.30	6.90
12.70	9.10	8.60	4.60	3.10	−0.70	0.40	−1.20	0.20	3.10	4.40	8.60
13.00	11.20	8.80	4.60	3.90	−0.20	0.60	−1.00	0.60	3.30	5.60	8.90
15.10	12.20	8.80	4.90	4.20	−0.10	0.70	−1.00	1.00	3.40	6.00	9.00
16.00	12.20	9.60	6.80	4.20	0.90	2.10	−0.70	1.00	3.80	7.20	9.60
16.00	12.60	10.10	6.80	4.50	1.40	3.20	−0.60	1.10	5.00	8.00	10.00
16.70	14.00	11.10	7.00	4.60	1.40	3.20	−0.10	1.40	5.60	8.20	10.50
16.80	14.10	11.60	7.20	4.60	1.60	3.30	−0.10	1.80	6.30	8.30	11.00
16.90	14.20	11.80	8.60	4.80	1.60	3.30	0.10	2.00	6.40	8.90	11.60
17.80	14.30	12.70	8.70	4.90	1.60	3.40	0.40	2.10	6.60	9.20	11.60

Table 3. Sample data for Y2020.

2020											
H1						H2					
Q1			Q2			Q3			Q4		
J	F	M	A	M	J	J	A	S	O	N	D
9.60	9.00	8.40	4.40	1.30	−2.70	−2.60	−3.70	−2.10	3.80	4.40	6.40
10.70	9.80	9.20	4.80	1.60	−2.40	−1.80	−1.90	−0.30	3.90	5.30	6.60
11.80	10.80	9.70	5.00	1.70	−1.00	−1.60	−1.30	0.40	4.00	7.00	6.60
12.30	10.90	9.90	5.20	2.20	−0.80	−1.20	−0.60	0.40	4.10	7.80	7.80
12.90	11.00	9.90	5.40	2.20	−0.70	−1.20	0.00	1.00	5.90	9.00	9.00
12.90	12.60	10.00	5.60	2.20	−0.60	−0.90	0.20	1.00	6.40	9.20	9.10
13.40	12.80	10.40	6.50	2.40	−0.10	−0.60	0.30	1.90	6.40	9.30	9.10
14.00	13.10	10.70	6.80	2.40	0.80	−0.30	0.60	2.50	6.50	9.50	9.40
14.00	13.90	11.00	7.30	2.80	1.10	−0.20	0.90	3.50	6.70	9.90	9.60
14.20	14.10	11.20	7.60	2.90	1.20	0.00	1.30	3.70	7.30	10.10	10.00

Table 4. WinDrift: driftFlags (dF) for Mode I & II.

(a) Mode I Drift Assessment Operations.

dataBlock	Mode I Operations	dF
1	No assessment	
2	Jan 2019 – Feb 2019	1
3	Feb 2019 – Mar 2019	0
4	Mar 2019 – Apr 2019	1
5	Apr 2019 – May 2019	0
6	May 2019 – Jun 2019;	1
	Q1 2019 and Q2 2019	1
7	Jun 2019 – Jul 2019	0
8	Jul 2019 – Aug 2019	1
9	Aug 2019 – Sep 2019;	1
	Q2 2019 and Q3 2019	1
10	Sep 2019 – Oct 2019	1
11	Oct 2019 – Nov 2019	0
12	Nov 2019 – Dec 2019;	1
	Q3 2019 and Q4 2019;	1
	H1 2019 and H2 2019	1

(b) Mode II Drift Assessment Operations.

dataBlock	Mode II Operations	dF
13	Jan 2019 – Jan 2020	1
14	Feb 2019 – Feb 2020	0
15	Mar 2019 – Mar 2020;	0
	Q1 2019 and Q1 2020	0
16	Apr 2019 – Apr 2020	0
17	May 2019 – May 2020	1
18	Jun 2019 – Jun 2020;	0
	Q2 2019 and Q2 2020;	0
	H1 2019 and H1 2020	0
19	Jul 2019 – Jul 2020	1
20	Aug 2019 – Aug 2020	0
21	Sep 2019 – Sep 2020;	0
	Q3 2019 and Q3 2020	0
22	Oct 2019 – Oct 2020	0
23	Nov 2019 – Nov 2020	0
24	Dec 2019 – Dec 2020;	0
	Q4 2019 and Q4 2020;	0
	H2 2019 and H2 2020	0

Drift assessment between two consecutive windows in Table 4a and between two corresponding windows in Table 4b.

of our design. Here, the dataset presents temperature data containing ten observations per month on an arbitrary location from Jan 2019–Dec 2020, as tabulated in Tables 2 and 3.

3.2 Step-by-Step Description

We start our discussion by presenting the complete algorithm of WinDrift as shown in Algorithm 1 . This is used to explain the steps involved in sample operations between as presented in Figs. 1c and 1d. The entire algorithm is divided into three steps:

- *Step 1: Get UserInput* – In this step, the user input is taken, and variables initialized.
- *Step 2: Prepare Windows* – Next, the sizes of the minimum, intermediate, and the maximum windows are determined to prepare for the drift assessment operation.

Table 5. Mode I: Data in windows.

Table 6. Mode I: ECDF of windows.

Table 7. Mode II: Data in windows.

Table 8. Mode II: ECDF of windows.

W_{H1} (Jan 2019)	W_{N1} (Feb 2019)
12.50	9.10
12.70	9.10
13.00	11.20
15.10	12.20
16.00	12.20
16.00	12.60
16.70	14.00
16.80	14.10
16.90	14.20
17.80	14.30

$\hat{F_{H1}}$ (Jan 2019)	$\hat{F_{N1}}$ (Feb 2019)
0.10	0.10
0.20	0.20
0.30	0.30
0.40	0.40
0.50	0.50
0.60	0.60
0.70	0.70
0.80	0.80
0.90	0.90
1.00	1.00

W_{H2} (Sep 2019)	W_{N2} (Sep 2020)
−0.30	−2.10
0.20	−0.30
0.60	0.40
1.00	0.40
1.00	1.00
1.10	1.00
1.40	1.90
1.80	2.50
2.00	3.50
2.10	3.70

$\hat{F_{H2}}$ (Sep 2019)	$\hat{F_{N2}}$ (Sep 2020)
0.10	0.10
0.20	0.20
0.30	0.30
0.40	0.40
0.50	0.50
0.60	0.60
0.70	0.70
0.80	0.80
0.90	0.90
1.00	1.00

- *Step 3: Detect Drift* – Finally, we assess drift on windows created above using functions explained in Sect. 3.1 i.e., *LevelCheck*, *SlidingWindow*, *CalculateECDF*, and *AssessDrift*. The results are reported through *driftFlags (dF)* on each level.

We provide a detailed explanation of these steps below:

Step 1: Get UserInput – In this step, the two types of inputs, i.e., mandatory and optional are taken to initialize the algorithm. The mandatory inputs include *dataBlocks* and *winLevSize* as explained in Sect. 3.1. The optional inputs include the α parameter as per Eq. 3. The default value of α is 0.050 to achieve a 95% confidence interval (CI); can also be altered by the user. The results of *dF* are reported continuously in the form of binary values, where, 1 represents *drift detected*, and 0 means *no drift*.

Step 2: Prepare Windows – Here, we prepare windows using the *winLevSize* array. The information is maintained on each level using unit size of a window i, indexes to monitor windows j, and count of cycles after *maxCycLen* is achieved i.e., m; ensuring that the drift is assessed between the correct windows. A generic expression of *dataBlocks* required for *Mode I* is (W_{H1}, W_{N1}) and *Mode II* is (W_{H2}, W_{N2}). As an example we selected *winLevSize* of $\{12, 6, 3, 1\}$ meaning that the first drift is assessed in *Mode I* when 2 *dataBlocks* arrive, and then in *Mode II* when 13 *dataBlocks* arrive i.e., one *dataBlocks* after the *maxCycLen* has reached at 12 *dataBlocks*.

Step 3: Detect Drift – In this step, we explain how WinDrift assesses drift into the two modes *Mode I* and *Mode II*. To assess drift in *Mode I* between (W_{H1}, W_{N1}) i.e., (Jan 2019, Feb 2019), we require only 2 *dataBlocks*; whereas, for *Mode II* between (W_{H2}, W_{N2}) i.e., (Sep 2019, Sep 2020), after Sep 2019, we wait for additional 12 *dataBlocks* to assess drift. The calculations are presented in the below sub-sections.

Step 3a: Assess Drift (Consecutive Mode) – Using the same example of *winLevSize* = $\{12, 6, 3, 1\}$, we present the expected drift operations for *Mode I* in Table 4a. Upon the arrival of *dataBlock* 2, the drift assessment begins only on a monthly level and continues until the year-end. Whereas, when *dataBlock* 6 arrives, only then the first quarterly assessment takes place, and the next assessments occur upon the arrival of *dataBlock* 9 and 12 respectively. Finally, the half-yearly assessment occurs when *dataBlock* 12 arrives. Note that the assessments on all window levels occur simultaneously. Once the year-end is reached, we automatically start assessing drift between two corresponding windows without continuing any further assessment between the consecutive windows.

As an example, we present drift assessment between (W_{H1}, W_{N1}) i.e., (Jan 2019, Feb 2019). For all other operations, the same process is repeated. Table 5 presents the sample observations taken from the first two columns of Table 2. In Table 6, the observed ECDFs $\hat{F}(x)$ for these months are presented. Next D_s is calculated using Eq. 2 which gives 0.700. D_c is calculated using Eq. 3. The windows contain 10 observations each for n_H and n_N and 0.050 is used as the default value of α. This gives a D_c of 0.607. Finally, the condition $D_s > D_c$ is assessed. Since $D_s > D_c$ then the drift is detected. This can be further validated through the visual inspection of Fig. 1c. Both results demonstrate that the drift is detected between the two windows.

Step 3b: Assess Drift (Corresponding Mode) – For the above-given example, we present a summary of expected operations for *Mode II* in Table 4b. The drift assessment on a monthly level starts when *dataBlock* 13 arrives and continues until the year-end. Whereas, as the 15th *dataBlock* arrives, the first quarterly assessment is performed followed by the assessments being carried out as *dataBlock* 18, 21, and 24 arrived. Similarly, for the first half-yearly assessment to occurs, we wait for the 18th *dataBlock* to arrive, followed by the next assessment at the year-end i.e., at the 24th *dataBlock*. Also, a yearly assessment between Y2019 and Y2020 occur at the year-end when two complete years' worth of data. Again, the assessments on all levels are carried out simultaneously.

For illustration, a sample drift assessment operation (W_{H2}, W_{N2}) i.e., (Sep 2019, Sep 2020), for *Mode II* is presented. The same process is repeated for all other operations. Table 7 presents the sample observations. These values are extracted from the ninth column of Tables 2 and 3 respectively. In Table 8, the observed ECDFs $\hat{F}(x)$ are presented. D_s, and D_c of both *datablocks* are calculated using the same method outlined in Step 3a. This gives us the same D_c as above, i.e., 0.607. As D_c is only dependent on n, and α, the value remains unchanged. However, the D_s is now 0.300. Since $D_s < D_c$, there is no drift detected; confirmed through the visual inspection of Fig. 1d.

4 Experimental Results

This section is divided into three parts as follows: Sect. 4.1 provides details on our experimental setup. In Sect. 4.2, we introduce the real and synthetic datasets used in this research. Section 4.3 presents our preliminary results in line with our objectives.

4.1 Experimental Setup

All the experiments are performed on a machine with 12 cores[1] and 32.0 GB of RAM. The coding is performed using PyCharm IDE 2021.2 [29]. We first establish the def-

---

[1] Intel(R) Xeon(R) CPU i7–9750H 2.60 GHz. System type 64–bit OS, x64-based processor.

Table 9. Statistical Properties used in generating Synthetic Datasets.

Data pattern	Stat char	Y2019 J	F	M	A	M	J	J	A	S	O	N	D	Y2020 J	F	M	A	M	J	J	A	S	O	N	D
C_1	μ_1	5.00	6.25	7.50	8.75	10.00	11.25	12.50	13.75	15.00	16.25	17.50	18.75	5.00	6.25	7.50	8.75	10.00	11.25	12.50	13.75	15.00	16.25	17.50	18.75
	σ_1	1.00	1.25	1.50	1.75	2.00	2.25	2.50	2.75	3.00	3.25	3.50	3.75	1.00	1.25	1.50	1.75	2.00	2.25	2.50	2.75	3.00	3.25	3.50	3.75
C_2	μ_2	4.90	6.15	7.40	8.65	9.90	11.15	12.40	13.65	14.90	16.15	17.40	18.65	4.90	6.15	7.40	8.65	9.90	11.15	12.40	13.65	14.90	16.15	17.40	18.65
	σ_2	0.90	1.15	1.40	1.65	1.90	2.15	2.40	2.65	2.90	3.15	3.40	3.65	0.90	1.15	1.40	1.65	1.90	2.15	2.40	2.65	2.90	3.15	3.40	3.65
C_3	μ_3	4.80	6.05	7.30	8.55	9.80	11.05	12.30	13.55	14.80	16.05	17.30	18.55	4.80	6.05	7.30	8.55	9.80	11.05	12.30	13.55	14.80	16.05	17.30	18.55
	σ_3	0.80	1.05	1.30	1.55	1.80	2.05	2.30	2.55	2.80	3.05	3.30	3.55	0.80	1.05	1.30	1.55	1.80	2.05	2.30	2.55	2.80	3.05	3.30	3.55
C_4	μ_4	5.10	6.35	7.60	8.85	10.10	11.35	12.60	13.85	15.10	16.35	17.60	18.85	5.10	6.35	7.60	8.85	10.10	11.35	12.60	13.85	15.10	16.35	17.60	18.85
	σ_4	1.10	1.35	1.60	1.85	2.10	2.35	2.60	2.85	3.10	3.35	3.60	3.85	1.10	1.35	1.60	1.85	2.10	2.35	2.60	2.85	3.10	3.35	3.60	3.85
C_5	μ_5	5.20	6.45	7.70	8.95	10.20	11.45	12.70	13.95	15.20	16.45	17.70	18.95	5.20	6.45	7.70	8.95	10.20	11.45	12.70	13.95	15.20	16.45	17.70	18.95
	σ_5	1.20	1.45	1.70	1.95	2.20	2.45	2.70	2.95	3.20	3.45	3.70	3.95	1.20	1.45	1.70	1.95	2.20	2.45	2.70	2.95	3.20	3.45	3.70	3.95
C_6	μ_6	5.00	6.25	7.50	8.75	10.00	11.25	12.50	13.75	15.00	16.25	17.50	18.75	18.75	17.50	16.25	15.00	13.75	12.50	11.25	10.00	8.75	7.50	6.25	5.00
	σ_6	1.00	1.25	1.50	1.75	2.00	2.25	2.50	2.75	3.00	3.25	3.50	3.75	3.75	3.50	3.25	3.00	2.75	2.50	2.25	2.00	1.75	1.50	1.25	1.00
C_7	μ_7	4.90	6.15	7.40	8.65	9.90	11.15	12.40	13.65	14.90	16.15	17.40	18.65	18.65	17.40	16.15	14.90	13.65	12.40	11.15	9.90	8.65	7.40	6.15	4.90
	σ_7	0.90	1.15	1.40	1.65	1.90	2.15	2.40	2.65	2.90	3.15	3.40	3.65	3.65	3.40	3.15	2.90	2.65	2.40	2.15	1.90	1.65	1.40	1.15	0.90
C_8	μ_8	4.80	6.05	7.30	8.55	9.80	11.05	12.30	13.55	14.80	16.05	17.30	18.55	18.55	17.30	16.05	14.80	13.55	12.30	11.05	9.80	8.55	7.30	6.05	4.80
	σ_8	0.80	1.05	1.30	1.55	1.80	2.05	2.30	2.55	2.80	3.05	3.30	3.55	3.55	3.30	3.05	2.80	2.55	2.30	2.05	1.80	1.55	1.30	1.05	0.80
C_9	μ_9	5.10	6.35	7.60	8.85	10.10	11.35	12.60	13.85	15.10	16.35	17.60	18.85	18.85	17.60	16.35	15.10	13.85	12.60	11.35	10.10	8.85	7.60	6.35	5.10
	σ_9	1.10	1.35	1.60	1.85	2.10	2.35	2.60	2.85	3.10	3.35	3.60	3.85	3.85	3.60	3.35	3.10	2.85	2.60	2.35	2.10	1.85	1.60	1.35	1.10
C_{10}	μ_{10}	5.20	6.45	7.70	8.95	10.20	11.45	12.70	13.95	15.20	16.45	17.70	18.95	18.95	17.70	16.45	15.20	13.95	12.70	11.45	10.20	8.95	7.70	6.45	5.20
	σ_{10}	1.20	1.45	1.70	1.95	2.20	2.45	2.70	2.95	3.20	3.45	3.70	3.95	3.95	3.70	3.45	3.20	2.95	2.70	2.45	2.20	1.95	1.70	1.45	1.20

inition of a drift and then compare our detection performance with 5 state-of-the-art drift detection methods. These methods include Adaptive WINdowing (ADWIN) [5], Early Drift Detection Method (EDDM) [4], Drift Detection Method based on Hoeffding's bounds with Moving Average-test ($HDDM_A$), and Drift Detection Method based on McDiarmid's bounds with Exponentially Weighted Moving Average (EWMA) ($HDDM_W$) [10], and Page Hinkley Test (PHT) [21]; these are configured with default parameter values as specified in Scikit multi-flow framework [19]. We also assess the performance of WD in terms of the computational resource utilization by using a *profiler* [23]. On an average, these experiments are run a 10 times for 21,600 observations.

4.2 Datasets

In this sub-section, we discuss the two types of datasets on which WinDrift is being tested. The synthetic datasets are designed first as they provide the known drift locations to evaluate the detection performance. An investigation with a real dataset confirms that WinDrift can detect drift in time-series obtained from a real-life application.

Synthetic Datasets: We simulate 2 experiments that will help us in validating our results based on the statistical characteristics of the datasets by randomly generating a normal distribution with certain mean μ and standard deviation σ for a single numerical attribute over the two consecutive years Y2019, and Y2020, as shown in Table 9. In *Experiment 1*, we create a similar sample distribution for both Y2019, and Y2020 expecting that there is *no drift* between their corresponding time periods. In *Experiment 2*, we create a dissimilar sample distribution for both Y2019, and Y2020 expecting that there is a *drift* between each of the corresponding time periods. These datasets follow a unique pattern such that C_1, C_2, C_3, C_4, and C_5 follow the objectives set out in *Experiment 1* while, C_6, C_7, C_8, C_9, and C_{10} are used in *Experiment 2*.

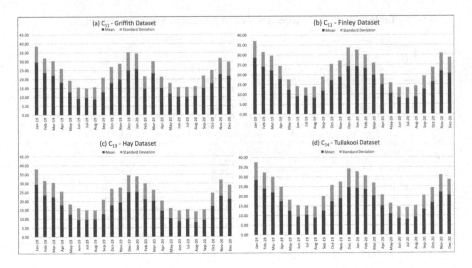

Fig. 2. Real Dataset from weather stations in Griffith, Finley, Hay, and Tullakool in Australia [7].

These values for each observation are derived from C_1 in Jan 2019, where we arbitrary select $\mu_1 = 5.00$ and $\sigma_1 = 1.00$. Thereafter, for the sake of consistency, month on month, we increase the value of μ by a constant 1.25 °C, and σ by a constant 0.25 °C with a minimum average temperature of 5 °C and a maximum average of 18.975 °C for C_1. Recall that the dataset C_6 is used in Experiment 2 so we inverse the statistical characteristics of the distribution i.e., the μ and σ for Jan 2019 is used for Dec 2020, the μ and σ for Feb 2019 is used for Nov 2020, and so on. This experiment provides information that the two sample distributions Y2019 and Y2020 are dissimilar and there is drift between them. To derive other datasets such as C_1, C_2,... C_9, C_{10}, we use $\mu_{\{2,7,4,9\}} = \mu_1 \pm 10\%$; and $\mu_{\{3,8,5,10\}} = \mu_1 \pm 20\%$. The justification for these experiments is that our results should follow the same pattern as that of the characteristics of this data.

Real Datasets: In the next set of experiments, we use four real datasets to evaluate our proposed method. These datasets are collected from weather stations of Griffith, Finley, Hay, and Tullakool in the Riverina regions of Australia [7] from Jan 2019–Dec 2020, also referred to as C_{11}, C_{12}, C_{13}, and C_{14}, respectively, in Sect. 4.3. These datasets contain three temperature observations over the 24 h of a day representing a minimum, average, and maximum temperature throughout the day, providing a large number of observation in a month. In an attempt to interpret the characteristics of this data, we create a graphical representation of all four datasets as shown in Fig. 2.

4.3 Numerical Results

As narrated in Sect. 1, there are two main objectives of this work: *Objective # 1* aims to achieve a *high detection performance* across hierarchical levels as presented in Table 10,

Table 10. DriftFlags (dFs) for WinDrift across multiple levels.

Experiments	Data pattern	Data nature	Expected results	Total dFs	L1 dFs	L2 dFs	L3 dFs	L4 dFs
1	C_1	Synthetic	ND	D = 3	D = 0	D = 0	D = 2	D = 1
				ND = 16	ND = 12	ND = 4	ND = 0	ND = 0
2	C_2	Synthetic	ND	D = 3	D = 0	D = 0	D = 2	D = 1
				ND = 16	ND = 12	ND = 4	ND = 0	ND = 0
3	C_3	Synthetic	ND	D = 4	D = 0	D = 1	D = 2	D = 1
				ND = 15	ND = 12	ND = 3	ND = 0	ND = 0
4	C_4	Synthetic	ND	D = 3	D = 0	D = 0	D = 2	D = 1
				ND = 16	ND = 12	ND = 4	ND = 0	ND = 0
5	C_5	Synthetic	ND	D = 3	D = 0	D = 0	D = 2	D = 1
				ND = 16	ND = 12	ND = 4	ND = 0	ND = 0
6	C_6	Synthetic	D	D = 19	D = 12	D = 4	D = 2	D = 1
				ND = 0	ND = 0	ND = 0	ND = 0	ND = 0
7	C_7	Synthetic	D	D = 18	D = 11	D = 4	D = 2	D = 1
				ND = 1	ND = 1	ND = 0	ND = 0	ND = 0
8	C_8	Synthetic	D	D = 18	D = 11	D = 4	D = 2	D = 1
				ND = 1	ND = 1	ND = 0	ND = 0	ND = 0
9	C_9	Synthetic	D	D = 19	D = 12	D = 4	D = 2	D = 1
				ND = 0	ND = 0	ND = 0	ND = 0	ND = 0
10	C_{10}	Synthetic	D	D = 19	D = 12	D = 4	D = 2	D = 1
				ND = 0	ND = 0	ND = 0	ND = 0	ND = 0
11	C_{11}	Real	MD	D = 9	D = 2	D = 4	D = 2	D = 1
				ND = 10	ND = 10	ND = 0	ND = 0	ND = 0
12	C_{12}	Real	MD	D = 12	D = 7	D = 2	D = 2	D = 1
				ND = 7	ND = 5	ND = 2	ND = 0	ND = 0
13	C_{13}	Real	MD	D = 17	D = 11	D = 3	D = 2	D = 1
				ND = 2	ND = 1	ND = 1	ND = 0	ND = 0
14	C_{14}	Real	MD	D = 13	D = 8	D = 2	D = 2	D = 1
				ND = 6	ND = 4	ND = 2	ND = 0	ND = 0

Synthetic = Synthetic Datasets (refer to Table 12). *Real* = Real Datasets (refer to Fig. 2). *D* = Drift; *ND* = No Drift. *MD* = Mixed Drift. *Total dFs* = total number of Drift Flags for a dataset. dFs collected on these levels mean: *L1 dFs* = Level 1 (e.g., monthly), *L2 dFs* = Level 2 (e.g., quarterly), *L3 dFs* = Level 3 (e.g., half-yearly), and *L4 dFs* = Level 1 (e.g., yearly).

and a comparison with the best-in-class detection methods in Table 11; *Objective # 2* is to achieve the *least use of computational resources* as shown in Table 12.

Objective # 1 – Drift Detection Performance: Our experiments are designed so that a *drift* event can be clearly defined. The expectation is that the our method detects *drift* with less *false alarms*. In Table 10, we present our results in the form of *driftFlags*.

For synthetic datasets, our drift detection performance results are presented in Table 10 and 11. Recall that in Experiment 1, we do not expect to observe a drift between the corresponding windows which is confirmed in our results. However, except ADWIN on one synthetic dataset, and EDDM on all datasets; none of the other methods have identified that the two sample distributions are actually similar and that *no drift* exists between them. While in Experiment 2, as expected, we observe a drift between all corresponding time periods. Note that all methods except EDDM detect drift. For all experiments, we used the same default parameter values as set by the Scikit-multi-flow framework therefore, the reason for non-detection by EDDM is unknown.

Table 11. Comparison of Drift Detection Performance of 5 Drift Detection Methods against WD.

Iteration	Expected results	Actual drift results over 6 drift detection methods					
		ADWIN	EDDM	HDDM$_A$	HDDM$_W$	PHT	WD
1	ND	D = 117	D = 0	D = 1912	D = 1139	D = 95	D = 3
			WG = 0	WG = 415	WG = 497	WG = 0	ND = 16
		×	✓	×	×	×	✓
2	ND	D = 109	D = 0	D = 1934	D = 1274	D = 93	D = 3
			WG = 0	WG = 341	WG = 531	WG = 0	ND = 16
		×	✓	×	×	×	✓
3	ND	D = 109	D = 0	D = 1875	D = 1400	D = 94	D = 4
			WG = 0	WG = 306	WG = 440	WG = 0	ND = 15
		×	✓	×	×	×	✓
4	ND	D = 100	D = 0	D = 2083	D = 1438	D = 100	D = 3
			WG = 0	WG = 351	WG = 466	WG = 0	ND = 16
		×	✓	×	×	×	✓
5	ND	D = 0	D = 0	D = 382	D = 243	D = 0	D = 3
			WG = 0	WG = 86	WG = 62	WG = 0	ND = 16
		✓	✓	×	×	✓	✓
6	D	D = 106	D = 0	D = 1688	D = 1085	D = 48	D = 19
			WG = 0	WG = 264	WG = 512	WG = 0	ND = 0
		✓	×	✓	✓	✓	✓
7	D	D = 110	D = 0	D = 1873	D = 1378	D = 53	D = 18
			WG = 0	WG = 415	WG = 497	WG = 0	ND = 1
		✓	×	✓	✓	✓	✓
8	D	D = 114	D = 0	D = 1912	D = 1324	D = 50	D = 18
			WG = 0	WG = 415	WG = 525	WG = 0	ND = 1
		✓	×	✓	✓	✓	✓
9	D	D = 106	D = 0	D = 1912	D = 1297	D = 53	D = 19
			WG = 0	WG = 415	WG = 377	WG = 0	ND = 0
		✓	×	✓	✓	✓	✓
10	D	D = 106	D = 0	D = 1912	D = 1297	D = 53	D = 19
			WG = 0	WG = 415	WG = 377	WG = 0	ND = 0
		✓	×	✓	✓	✓	✓
11	MD	D = 34	D = 0	D = 326	D = 152	D = 87	D = 9
			WG = 0	WG = 251	WG = 284	WG = 0	ND = 10
		✓	×	✓	✓	✓	✓
12	MD	D = 54	D = 0	D = 552	D = 200	D = 128	D = 12
			WG = 0	WG = 382	WG = 426	WG = 0	ND = 7
		✓	×	✓	✓	✓	✓
13	MD	D = 65	D = 0	D = 598	D = 233	D = 141	D = 17
			WG = 0	WG = 483	WG = 468	WG = 0	ND = 2
		✓	×	✓	✓	✓	✓
14	MD	D = 53	D = 0	D = 562	D = 239	D = 134	D = 13
			WG = 0	WG = 381	WG = 382	WG = 0	ND = 6
		✓	×	✓	✓	✓	✓

D = Drift; ND = No Drift. MD = Mixed Drift. WG = Warning. ✓ = correct detection. × = incorrect detection.

For real datasets, some time periods contain *drift*; we call this *mixed drift*. As the exact drift points are not known; we apply the two-sample test using statistical distance [9] to determine the estimated location of the drift. A majority vote across the results received from the Cramér-von Mises (CvM) family of tests [6] provides an independent

Table 12. Estimated resource utilization.

Methods	Real datasets			Synthetic datasets		
	Runtime (sec)	CPU (%)	RAM (MB)	Runtime (sec)	CPU (%)	RAM (MB)
ADWIN	3.115	19.25	144.25	3.265	20.50	147.00
EDDM	**1.655**	17.25	135.50	1.550	**17.00**	136.00
$HDDM_A$	2.195	18.88	141.00	2.235	18.00	143.50
$HDDM_W$	2.373	18.75	139.50	2.365	17.50	140.50
PHT	1.880	20.50	135.50	1.745	22.00	137.50
WinDrift	**1.655**	**17.13**	**134.25**	**1.545**	**17.00**	**132.50**

The best result is given in boldface. Runtime is given in *sec*, % for the CPU utilization, and *MB* for the utilization of RAM.

baseline [17]. We also conducted a visual inspection of these datasets to confirm the time periods for *drift* and *no drift* events. We explain this by selecting examples from $C_{11},...C_{14}$. In C_{11}, WinDrift detects *drift* between Feb 2019 and Feb 2020, a visual inspection of Fig. 2 confirms that there is approx. 9–10°C difference in the average temperature of these two months. Another example is from C_{12} demonstrating that there is *no drift* between Sep 2019 and Sep 2020 as their characteristics appear identical. Similarly in C_{13} and C_{14}, the drift is identified in all months except between June 2019 and June 2020. All other results can be generated by repeating this process.

Objective # 2 – Computational Resources: We evaluate the performance of WinDrift on the estimated use of computational resources against the best-in-class drift detection methods. For synthetic datasets, across both experiments, we averaged the performance of the detection methods. The results of these preliminary investigations are presented in Table 12. Note that across both datasets, WD has the shortest runtime and consumes the least resources. We argue that the justification for this is the use of ECDF as sufficient statistics and the use of adaptive forgetting strategy. Instead of remembering all observations, we calculate the ECDF of $\hat{F}(x)$ of each distribution. We require only a single point in both distributions where their difference D_s is more than the threshold D_c. Soon after, any *irrelevant* data e.g., when assessing drift in Jan 2020 the data from Dec 2019 is not required; and *stale* data, e.g., when assessing Feb 2020, the data from Jan 2019 is not required is forgotten using the *winLevSize*. These characteristics make WD efficient and lightweight to execute in any incremental learning algorithm.

5 Conclusion and Future Work

Due to the emerging concepts in the real-world applications, changes may emerge in the data patterns over time. An in-depth analysis of time-series produced by such applications enabled us in developing a drift detection method that presents a strong case for its effectiveness. Our multiple time-based hierarchical windows allow the users to decide the granularity of drift detection they require for their applications. WinDrift primarily detects drift between the corresponding time periods both accurately and efficiently. The thoughtfulness of our method to consider real-life applications and user flexibility for our design, to the best of our knowledge, is the first stepping stone in this direction.

In our future work, we plan to investigate drift assessment functions, and evaluate the performance of our method in the presence of noise, and datasets with mixed drift.

References

1. Alippi, C., Boracchi, G., Roveri, M.: Hierarchical change-detection tests. IEEE Trans. Neural Networks Learn. Syst. **28**(2), 246–258 (2016)
2. Alippi, C., Roveri, M.: Just-in-time adaptive classifiers-part i: detecting nonstationary changes. IEEE Trans. Neural Networks **19**(7), 1145–1153 (2008)
3. Alippi, C., Roveri, M.: Just-in-time adaptive classifiers-part ii: designing the classifier. IEEE Trans. Neural Networks **19**(12), 2053–2064 (2008)
4. Baena-Garcıa, M., del Campo-Ávila, J., Fidalgo, R., Bifet, A., Gavalda, R., Morales-Bueno, R.: Early drift detection method. In: Fourth International Workshop on Knowledge Discovery from Data Streams, vol. 6, pp. 77–86 (2006)
5. Bifet, A., Gavaldà, R.: Learning from time-changing data with adaptive windowing. In: Proceedings of the Seventh SIAM International Conference on Data Mining, pp. 135–150. SIAM (2007)
6. Choulakian, V., Lockhart, R.A., Stephens, M.A.: Cramér-von mises statistics for discrete distributions. Can. J. Stat./La Rev. Can. Statistique, **22**, 125–137 (1994)
7. CSIRO: agriculture flagship. weather stations in Riverina (2021). https://weather.csiro.au/
8. Dasu, T., Krishnan, S., Venkatasubramanian, S., Yi, K.: An information-theoretic approach to detecting changes in multi-dimensional data streams. In: Proceedings Symposium on the Interface of Statistics, Computing Science, and Applications. Citeseer (2006)
9. Elmore, K.L.: Alternatives to the chi-square test for evaluating rank histograms from ensemble forecasts. Weather Forecast. **20**(5), 789–795 (2005)
10. Frias-Blanco, I., del Campo-Ávila, J., Ramos-Jimenez, G., Morales-Bueno, R., Ortiz-Diaz, A., Caballero-Mota, Y.: Online and non-parametric drift detection methods based on Hoeffding's bounds. IEEE Trans. Knowl. Data Eng. **27**(3), 810–823 (2014)
11. Gama, J.a., Žliobait, I., Bifet, A., Pechenizkiy, M., Bouchachia, A.: A survey on concept drift adaptation. ACM Comput. Surv. (CSUR), **46**(4), 1–37 (2014)
12. Gama, J., Medas, P., Castillo, G., Rodrigues, P.: Learning with drift detection. In: Bazzan, A.L.C., Labidi, S. (eds.) SBIA 2004. LNCS (LNAI), vol. 3171, pp. 286–295. Springer, Heidelberg (2004). https://doi.org/10.1007/978-3-540-28645-5_29
13. Habimana, J.R.: Analysis of break-points in financial time series. University of Arkansas (2016)
14. Kifer, D., Ben-David, S., Gehrke, J.: Detecting change in data streams. In: VLDB, vol. 4, pp. 180–191. Toronto, Canada (2004)
15. Liu, A.: Concept drift adaptation for learning with streaming data. Ph.D. thesis (2018)
16. Lu, N., Zhang, G., Lu, J.: Concept drift detection via competence models. Artif. Intell. **209**, 11–28 (2014)
17. Martínez-Camblor, P., Carleos, C., Corral, N.: Cramér-von mises statistic for repeated measures. Revista Colombiana de Estadística **37**(1), 45–67 (2014)
18. Mehmood, H., Kostakos, P., Cortes, M., Anagnostopoulos, T., Pirttikangas, S., Gilman, E.: Concept drift adaptation techniques in distributed environment for real-world data streams. Smart Cities **4**(1), 349–371 (2021)
19. Montiel, J., Read, J., Bifet, A., Abdessalem, T.: Scikit-multiflow: A multi-output streaming framework. J. Mach. Learn. Res. **19**(1), 2915–2914 (2018)
20. Naqvi, N., Rehman, S.U., Islam, M.Z.: A hyperconnected smart city framework: digital resources using enhanced pedagogical techniques. Australas. J. Inf. Syst. 24 (2020)

21. Page, E.S.: Continuous inspection schemes. Biometrika **41**(1/2), 100–115 (1954)
22. Pratt, J.W., Gibbons, J.D.: Kolmogorov-Smirnov two-sample tests. In: Concepts of Nonparametric Theory. Springer Series in Statistics, pp. 318–344. Springer, New York (1981). https://doi.org/10.1007/978-1-4612-5931-2_7
23. PyPI: psutil 5.8.0 (2020). https://pypi.org/project/psutil/
24. Rahman, M.G., Islam, M.Z.: Adaptive decision forest: an incremental machine learning framework. Pattern Recogn. **122**, 108345 (2022)
25. Raza, H., Prasad, G., Li, Y.: EWMA model based shift-detection methods for detecting covariate shifts in non-stationary environments. Pattern Recogn. **48**(3), 659–669 (2015)
26. Shao, J., Ahmadi, Z., Kramer, S.: Prototype-based learning on concept-drifting data streams. In: Proceedings of the 20th ACM SIGKDD International Conference on Knowledge Discovery and Data Mining, pp. 412–421 (2014)
27. Siegel, S., Castellan, N.: 2nd edition: Nonparametric statistics for the behavioral sciences (1988)
28. Van der Vaart, A.W.: Asymptotic Statistics, vol. 3. Cambridge University Press, Cambridge (2000)
29. Van Rossum, G., Drake Jr, F.L.: Python reference manual. Centrum voor Wiskunde en Informatica Amsterdam (1995)
30. Wang, H., Abraham, Z.: Concept drift detection for streaming data. In: 2015 International Joint Conference on Neural Networks (IJCNN), pp. 1–9. IEEE (2015)
31. Watada, J.: Kolmogorov-Smirnov two sample test with continuous fuzzy data, pp. 175–186 (2010)
32. Webb, G.I., Lee, L.K., Goethals, B., Petitjean, F.: Analyzing concept drift and shift from sample data. Data Min. Knowl. Discov. **32**(5), 1179–1199 (2018). https://doi.org/10.1007/s10618-018-0554-1
33. Widmer, G., Kubat, M.: Learning in the presence of concept drift and hidden contexts. Mach. Learn. **23**, 69–101 (1996). https://doi.org/10.1023/A:1018046501280
34. Yu, S., Abraham, Z.: Concept drift detection with hierarchical hypothesis testing. In: Proceedings of the 2017 SIAM International Conference on Data Mining, pp. 768–776. SIAM (2017)

Investigation of Explainability Techniques for Multimodal Transformers

Krithik Ramesh[✉] and Yun Sing Koh

School of Computer Science, University of Auckland, Auckland, New Zealand
kram416@aucklanduni.ac.nz, y.koh@auckland.ac.nz

Abstract. Multimodal transformers such as CLIP and ViLBERT have become increasingly popular for visiolinguistic tasks as they have an efficient and generalizable understanding of visual features and labels. Notable examples of visiolinguistic models include OpenAI's CLIP by Radford et al. and ViLBERT by Lu et al. One of the gaps in current multimodal transformers is that there are no unified explainability frameworks to compare attention interactions meaningfully between models. To address the comparability concern, we investigate two different explainability frameworks. Specifically, Label Attribution and Optimal Transport of Vision-Language semantic spaces with the Visual-BERT multimodal transformer model provide an interpretability process towards understanding attention interactions in multimodal transformers. We provide a case study of the Visual Genome and Question Answer 2 Datasets trained using VisualBERT.

Keywords: Multimodal transformers · Label attribution · Optimal transport

1 Introduction

Multimodal transformers are commonplace for visiolinguistic tasks such as video search, image captioning, and product recommendation. Multimodal transformers, such as OpenAI's CLIP [19], and Facebook AI Research's ViLBERT [17] have achieved higher accuracy on classification tasks after fewer training iterations compared to pure natural language processing (NLP) or computer vision (CV) models. They combine learned information from both modalities to understand the problem space. However, the nature of this multimodal learning is difficult to interpret. The motivation for this research is that fine-tuning multimodal transformer models like VisualBERT [15] poses some practical obstacles, such as over-biased embeddings or a lack of generalized learning for a given task [2,5].

Current methods employed to interpret these interactions focus on one component of multimodal transformers. In the case of Nielson et al. [18] and Schmidht et al. [20], the explainability of visual and linguistic pairings is understood by using sequencing techniques more suitable for pure text input. VisualBERT, developed by Li et al. [15], uses the implicit alignment of text embeddings

L. A. F. Park et al. (Eds.): AusDM 2022, CCIS 1741, pp. 90–98, 2022.
https://doi.org/10.1007/978-981-19-8746-5_7

with visual features and measured raw gradient attributions. From a statistical perspective, Wasserstein Distances is a common technique used to quantify semantic relationships inside transformer embeddings [13]. A range of implementations to understand embedding spaces, such as topic modeling and distillation techniques, have been proposed [21,26].

Substantial work has been carried out in specific components of explainability and domain adaption for transformer and computer vision models. However, the emergence of multimodal vision and text-based models is relatively recent [1]. As multimodal transformers are in their infancy, explainability techniques used to analyze these models generally address limited modalities or do not present information about embedding feature modifications [9]. The two main papers that lay the foundation for this research are the generic bi-modal attention explainability by Chefer et al. [3] and Wasserstein distillation for understanding semantic relationships by Xu et al. [26].

We take a different perspective on multimodal explainability techniques by experimentally applying label attribution, and Wasserstein distances on VisualBERT to observe attribution and distributional phenomena between image and text embeddings, as shown in Sect. 3. Our main contributions are: First, we investigate attribution through a modified multimodal approach of the Visual Question Answer 2 dataset queries through VisualBERT. Second, we explore an image-text embedding space inside VisualBERT using Wasserstein distances.

2   Problem Definition

Multimodal transformers have complex interactions between image and text inputs as they are derived from different source distributions before embedding. We follow a similar formulation by Chefer et al. [4].

We define the image and text tokens as t, i, respectively. We use the symbols (t, i) jointly to express the association between the two domains. There are four different types of attention interactions between input tokens: The first two are self-attention interactions $A^{t,t}$ and $A^{i,i}$ for the text and image tokens, respectively. Then there are $A^{t,i}$, which represents the effect of text tokens on each image token, and $A^{i,t}$, which represents the effect of image tokens on each of the text tokens. However, there currently does not exist a standardized process to evaluate these attention interactions between different multimodal transformer architectures [6]. We address this problem by investigating image-text attention interactions using the aforementioned two explainability techniques on Visual-BERT.

We follow the notation of Xu et al. [26] for describing the composition of the word and visual embeddings. Let x_n be the embedding of the n-th word in $n \in \{1, \dots, N\}$ and u_n is the embedding of the u-th visual feature in $u \in \{1, \dots, U\}$. Such that we get $X = [x_n] \in \mathbb{R}^{D \cdot N}, U = [u_n] \in \mathbb{R}^{D \cdot U}$, where D denotes the distance between modalities. We provide the modifications and details about semantic information shared between modalities in Sect. 3.2.

2.1   Quantifying Syntactic Grounding Through Label Attribution

We aim to investigate the contextual relationship being inferred by VisualBERT between the text and image distributions. We leverage a well-understood approach called label-attribution [3] with a modification to the calculation of relevant information between layers using the gradient information between VisualBERTs Layer heads and then applying a Hadamard matrix, which we will discuss in Sect. 3.1. Label attribution allows us to associate attention from text queries to specific parts of images.

2.2   Investigating Semantic Relationships Through Optimal Transport

Here, we explain the need for statistical evaluation of semantic relations between image and text embeddings. Since self-attention models such as visualBERT embed images and text together, the context between the modalities is derived from dot product calculation to find alignment scores, followed by normalization of the weights and reweighing the original embeddings. However, the semantic relationship derived from these calculations is not immediately available with attention scores. To quantify the semantic relationship between text and images, we use t-stochastic neighbor embedding of image and text embeddings to develop lower-dimensional embedding clusters. We then use Wasserstein distances to calculate distances between the distributions for training VisualBERT on the Visual Genome Dataset. The optimal transport process is explained in detail in Sect. 3.2.

 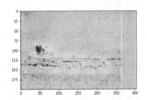 

(a) Original Image from VQA2 Dataset

(b) FPN Visual Embedding from P_2 Layer

(c) FPN Visual Embedding from P_3 Layer

Fig. 1. Visual embedding feature extraction from VQA2 using Detectron2. The scales on the images represent the pixel width and height.

3   Explainability Techniques

In this section, we describe two datasets, Visual Genome (VG) [11] and Visual Question Answer 2 (VQA2) [24], used to investigate the semantic relationships between text and image modalities represented as text and visual embeddings.

VisualBERT was trained on the Visual Genome dataset as it provides a breadth of structured and contextualized image descriptions. The VQA2 dataset provided a more semantically nuanced training and validation set as the questions provided some contextual ambiguity [7,24]. We construct these visual embeddings by extracting regions of interest from images in the VG and VQA2 datasets and then processing them through the Detectron2 [25] backbone and extracting the Feature Pyramid Network (FPN) layers $\{P_2, \ldots, P_5\}$ from Detectron2. These layers $\{P_2, \ldots, P_5\}$ represent the feature maps produced at each level of the FPN and are associated with their respective residual convolutions $\{C_2, \ldots, C_5\}$ [16]. For example, Fig. 1 shows a sample image followed by features extracted from FPN Layers P_2 and P_3.

While many multimodal transformer architectures exist, this work leverages VisualBERT as the multimodal transformer of interest. VisualBERT is a self-attention-based multimodal transformer, as opposed to co-attention-based models. We distinguish the differences between these models based on how the visual embeddings get presented in the transformer. For example, in the case of VisualBERT, as depicted in Fig. 2(c), the text and image embeddings are presented as one entity. Whereas co-attention models introduce visual embeddings as independent queries from the text embeddings, as seen in Figs. 2(a) and 2(b) respectively.

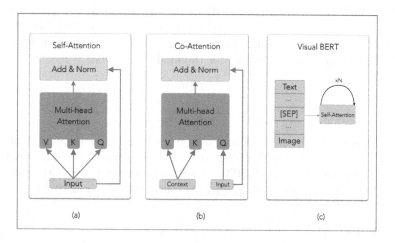

Fig. 2. Representation of the self-attention architecture (a), Representation of the self-attention architecture (b), overview of VisualBERT's self-attention mechanism's organization of image and text embeddings (c) [4].

We quantify the interactions that yield attention scores from visual and text embeddings using a modified layer-wise relevance propagation (LRP). This method is intended as a post-hoc gauge to assess the quality of the training and task performance of multimodal transformers. In the next section, we discuss the usage and modification of LRP for label attribution.

3.1   Label Attribution

This section explains the modifications performed on LRP to more explicitly evaluate syntactic grounding between image and text modalities through label attribution. LRP is considered a generally effective method of explaining transformers as it considers relevancy by each head in a layer instead of the mean attention for each layer [18]. However, LRP is limited in its ability to extract relevancy as it cannot explicitly distinguish between positive and negative attributions [4,18]. Meaning that high negative and positive attributions will cancel out, limiting the overall explainability of transformer models. Chefer et al. [3] propose a modification to LRP such that they can attribute attention with gradient and relevancy information non-linearly between one layer and the next.

LRP Modification. We again follow a similar definition for relevancy maps by Chefer et al. [4]. Derived from the structure of the attention interactions, we define a relevancy map with the same set of interactions. For self-attention, R^{tt}, R^{ii}, and R^{ti}, R^{it} for text-image attention. To update the relevancy information, the proposed method uses attention maps where each layer's heads are averaged using gradients. Let $\bar{A} \in \mathbb{E}_h(\nabla A \odot A)$, where $\odot$ defines the Hadamard product to update the attentions.

Here we follow the LRP proposed by Chefer et al. [3] and adapt it to Visual BERT's self-attention architecture. We define the parameters of a self-attention encoder-decoder transformer as proposed by Vaswani et al. [23]. Let O be the output of the attention model, Q be the queries matrix, and K, V the key, values matrices. We define these matrices for the model such that:

$$O \in \mathbb{R}^{h \cdot s \cdot d_h} \tag{1}$$

$$Q \in \mathbb{R}^{h \cdot s \cdot d_h}, K, V \in \mathbb{R}^{h \cdot q \cdot d_h} \tag{2}$$

where h denotes the number of heads, d_h represents the embedding dimension, and $s, q \in i, t$ indicates both the domain and the number of tokens within a domain. In the case of VisualBERT's self-attention architecture, attention takes place between q key tokens and s query tokens where $s = q$ [3,10,23].

3.2   Optimal Transport

This section explains the modifications made to the optimal transport implementation to calculate distances between the image and text modalities. After training on Visual Genome, we attempt to quantify the changes in the semantic feature space inside VisualBERT [14]. By modifying the data representation of embeddings to account for the dimensionality and distributional differences of independent text and image modalities, we extend the work of Xu et al. [26], Sun et al. [22], and Li et al. [14]

Wasserstein Topic Model Based on Euclidean Word Embeddings. We define M as the number of descriptions and N, V is the number of words and visual features in a given corpus [26]. These descriptions can be characterized as

$$Y = [y_m] \in \mathbb{R}^{N \cdot M \cdot V} \tag{3}$$

where y_m can be considered as the accumulation of the visual and text features of a given description [21, 26]:

$$y_m \in \Sigma^{N \cdot V}, m \in \{1, \dots, M\}. \tag{4}$$

Finally, we define any two features $i \in N$ and $z \in V$ such that the discrete distance between them is denoted as $D \in \mathbb{R}^{N^2 \cdot M^2}$ [22, 26]. Finally, we define T as the optimal transport matrix, in this model we assume that the distance between two words is the distance between embeddings [12, 14, 26]. Further details on the Wasserstein Distance calculations are shown in Xu et al. [26].

4 A Case Study in VisualBERT Explainability

We extend our explainability of VisualBERT by understating the label attribution assigned to each token in a query compared against the accuracy of the prediction. As previously mentioned, relevancy maps are not standardized nor immediately comparable as they are specific to the model architecture, as such, the results of the modified LRP and label attribution are mostly qualitative. Figure 3 explore randomly selected label attributions for queries with incorrect responses. Qualitatively, we recognize that longer descriptions or phrases yield high attribution to specific words, which sometimes distorts the intended meaning and leads to misclassification errors [8]. We find this attribution behavior is also consistent among self-attention multimodal transformers. In the third example in Fig. 3, we note that VisualBERT misidentified tokens as less relevant. Tokens `written` and `side` are not attributed heavily in the classification and contributed against the correct prediction label.

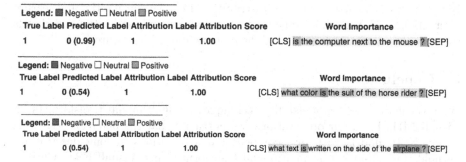

Fig. 3. Randomly visualized label attribution from visual genome and VQA2 questions.

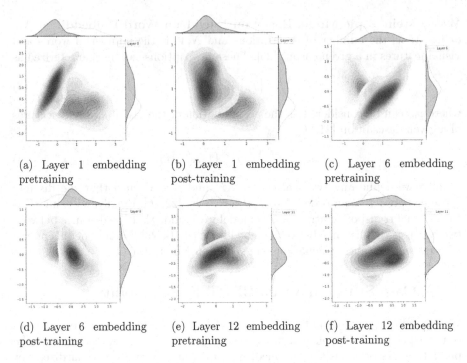

(a) Layer 1 embedding pretraining   (b) Layer 1 embedding post-training   (c) Layer 6 embedding pretraining

(d) Layer 6 embedding post-training   (e) Layer 12 embedding pretraining   (f) Layer 12 embedding post-training

Fig. 4. Wasserstein distributions of image and text embeddings produced by Visual-BERT before and after training on Visual Genome dataset. (Color figure online)

Here, we visualize the differences in the embedding space of VisualBERT after 25 generations of training on the Visual Genome dataset. In Fig. 4, we can see that the vision embeddings (blue) and text embeddings (red) begin to aggregate. This indicates the different modalities of the same semantic feature are beginning to aggregate. This behavior is analogous to the baseline testing by Singh et al. [21] and Xu et al. [26] and the distribution of features begin to overlap and the distance between the Barycenters proportionally. The interpretations are limited as out-of-domain testing would describe semantics in more detail. However, as far as understanding distributional changes in multimodal feature interactions, this provides a quantitative surface leveling understanding that is comparable to existing baselines in the aforementioned literature.

5   Conclusion

We investigated two explainability techniques for multimodal transformers on VisualBERT. From our observations, we postulate that the behavior seen with purely NLP models with over-compensated attributions in long descriptions [8] holds for self-attention multimodal transformers like VisualBERT. Current research [8,24] suggests that as the number of tokens increases, the probability for specific tokens, generally esoteric terms, gets weighted heavily. While image tokens provide more contextual information to VisualBERT, the features extracted by Detectron2 may provide distorted information to answer a given question.

This is an initial attempt at the experimental usage of explainability techniques for multimodal transformers. Future considerations include attempting to generalize these explainability techniques to all visiolinguistic multimodal transformers irrespective of the attention mechanism. Another potential consideration is investigating the vision and text embeddings more hierarchically using visual scene graphs and label attribution.

References

1. Baltrusaitis, T., Ahuja, C., Morency, L.P.: Multimodal machine learning: a survey and taxonomy. IEEE Trans. Pattern Anal. Mach. Intell. **41**(2), 423–443 (2019). https://doi.org/10.1109/TPAMI.2018.2798607
2. Brunner, G., Liu, Y., Pascual, D., Richter, O., Wattenhofer, R.: On the validity of self-attention as explanation in transformer models. CoRR abs/1908.04211 (2019). https://arxiv.org/abs/1908.04211
3. Chefer, H., Gur, S., Wolf, L.: Generic attention-model explainability for interpreting bi-modal and encoder-decoder transformers. In: Proceedings of the IEEE/CVF International Conference on Computer Vision, pp. 397–406 (2021)
4. Chefer, H., Gur, S., Wolf, L.: Transformer interpretability beyond attention visualization. In: Proceedings of the IEEE/CVF Conference on Computer Vision and Pattern Recognition, pp. 782–791 (2021)
5. Clark, K., Khandelwal, U., Levy, O., Manning, C.D.: What does BERT look at? An analysis of BERT's attention. In: BlackBoxNLP@ACL (2019)
6. Devlin, J., Chang, M.W., Lee, K., Toutanova, K.: Bert: pre-training of deep bidirectional transformers for language understanding. arXiv preprint arXiv:1810.04805 (2018)
7. Gao, C., Zhu, Q., Wang, P., Wu, Q.: Chop chop BERT: visual question answering by chopping VisualBERT's heads (2021)
8. Janizek, J.D., Sturmfels, P., Lee, S.I.: Explaining explanations: axiomatic feature interactions for deep networks. J. Mach. Learn. Res. **22**(104), 1–54 (2021). https://jmlr.org/papers/v22/20-1223.html
9. Joshi, G., Walambe, R., Kotecha, K.: A review on explainability in multimodal deep neural nets. IEEE Access **9**, 59800–59821 (2021)
10. Kiela, D., Bhooshan, S., Firooz, H., Perez, E., Testuggine, D.: Supervised multimodal bitransformers for classifying images and text. arXiv preprint arXiv:1909.02950 (2019)
11. Krishna, R., et al.: Visual genome: connecting language and vision using crowdsourced dense image annotations (2016)
12. Kusner, M., Sun, Y., Kolkin, N., Weinberger, K.: From word embeddings to document distances. In: International Conference on Machine Learning, pp. 957–966. PMLR (2015)
13. Lee, J., Dabagia, M., Dyer, E., Rozell, C.: Hierarchical optimal transport for multimodal distribution alignment. In: Advances in Neural Information Processing Systems, vol. 32 (2019)

14. Li, C., Li, X., Ouyang, J., Wang, Y.: Semantics-assisted Wasserstein learning for topic and word embeddings. In: 2020 IEEE International Conference on Data Mining (ICDM), pp. 292–301 (2020). https://doi.org/10.1109/ICDM50108.2020.00038

15. Li, L.H., Yatskar, M., Yin, D., Hsieh, C.J., Chang, K.W.: VISUALBERT: a simple and performant baseline for vision and language. arXiv preprint arXiv:1908.03557 (2019)

16. Lin, T.Y., Dollár, P., Girshick, R., He, K., Hariharan, B., Belongie, S.: Feature pyramid networks for object detection. In: Proceedings of the IEEE Conference on Computer Vision and Pattern Recognition, pp. 2117–2125 (2017)

17. Lu, J., Batra, D., Parikh, D., Lee, S.: ViLBERT: pretraining task-agnostic visiolinguistic representations for vision-and-language tasks. In: Advances in Neural Information Processing Systems, vol. 32 (2019)

18. Nielsen, I.E., Dera, D., Rasool, G., Bouaynaya, N., Ramachandran, R.P.: Robust explainability: a tutorial on gradient-based attribution methods for deep neural networks. arXiv preprint arXiv:2107.11400 (2021)

19. Radford, A., et al.: Learning transferable visual models from natural language supervision. In: International Conference on Machine Learning, pp. 8748–8763. PMLR (2021)

20. Schmidt, F., Hofmann, T.: Bert as a teacher: contextual embeddings for sequence-level reward. arXiv preprint arXiv:2003.02738 (2020)

21. Singh, S.P., Hug, A., Dieuleveut, A., Jaggi, M.: Context mover's distance & barycenters: optimal transport of contexts for building representations (2020)

22. Sun, C., Yan, H., Qiu, X., Huang, X.: Gaussian word embedding with a wasserstein distance loss. arXiv preprint arXiv:1808.07016 (2018)

23. Vaswani, A., et al.: Attention is all you need. In: Advances in Neural Information Processing Systems, vol. 30 (2017)

24. Wu, B., et al.: Visual transformers: token-based image representation and processing for computer vision (2020)

25. Wu, Y., Kirillov, A., Massa, F., Lo, W.Y., Girshick, R.: Detectron2 (2019). https://github.com/facebookresearch/detectron2

26. Xu, H., Wang, W., Liu, W., Carin, L.: Distilled Wasserstein learning for word embedding and topic modeling (2018)

Effective Imbalance Learning Utilizing Informative Data

Han Tai[1], Raymond Wong[1(✉)], and Bing Li[2]

[1] School of Computer Science and Engineering, University of New South Wales,
Sydney, Australia
han.tai@student.unsw.edu.au, ray.wong@unsw.edu.au
[2] The A*STAR Centre for Frontier AI Research (CFAR), Singapore, Singapore
li_bing@ihpc.a-star.edu.sg

Abstract. The class imbalance problem that is caused by unequal data distribution usually results in poor performance, and it has attracted increasing attention in the research community. The challenge of the problem is the difficulty to extract sufficient information from the minority class. As a result, the classifier converges to a sub-optimal state. While methods based on resampling and reweighing the cost for different classes are the common strategies to address the problem, there are still numerous issues with these methods such as under- or over-sampling that may remove necessary information or introduce noise, respectively; and reweighing may result in an inappropriate cost matrix.

To address the above shortcomings, in this paper, an enhanced approach based on informative samples is proposed. In our approach, the classifier can indicate which class a sample is closer to by comparing it with boundary samples. The informative samples include the samples from both positive and negative samples located around the boundary. Finally, our experiments show that our proposed method outperforms state-of-the-art algorithms by 18% on F_1 score.

Keywords: Imbalance learning · Informative data

1 Introduction

With the evolution of technology, an increasing amount of data is available. They provide the necessary information to support the developments of solutions to research problems in different domains, for example, node classification [45]. To solve classification problems, common classification algorithms generally assume the data in different classes are balanced [19]. Class imbalance problem is a special yet common case for classification problem where the number of instances of classes is not uniformly distributed [27]. Class imbalance can be found in a number of situations where collecting minority samples is extremely expensive or inherently rare, for instance, patients of a rare disease are hard to find. Other class imbalance applications includes recognizing objects and actions in video sequences (binary and multi-class problem) [29], abnormal detection [32], risk behavior recognition [9], etc.

Learning under class imbalance has attracted a significant number of interest in both academic and industrial areas [1,43]. Considering common classifiers work under the

© The Author(s), under exclusive license to Springer Nature Singapore Pte Ltd. 2022
L. A. F. Park et al. (Eds.): AusDM 2022, CCIS 1741, pp. 99–114, 2022.
https://doi.org/10.1007/978-981-19-8746-5_8

balanced-class presumption [19,42], the class imbalance makes the learning process lean towards the majority rather than the minority, pushing the decision boundary to the minority side, as a consequence, the classifier will blindly tag more majority class, leading to biased results [38]. Moreover, coming with the scarcity of minority data, the data overlapping and noise issues are exacerbated, which greatly deteriorates the classification results [22].

Owing to its great usefulness, a number of methods have been proposed to alleviate the effect of imbalanced-class, and all these methods can be broadly classified into two major categories, data-based solutions and algorithm-based solutions [41]. The strategy implemented by most data level methods is modifying the data distribution to a balanced state, which can be achieved by under-sampling, selecting samples from majority class [33], or over-sampling, generating new instances for minority samples [7]. Reducing the cost by assigning weights for samples belonging to different classes is the main idea for some of the recently proposed algorithms [34]. However, there are different drawbacks to these strategies. As the informative samples may be eliminated by the undersampling process, the amount of total information can be reduced, which can lead to a low precision [21]. In addition, the generated samples in oversampling methods can be the noise data that are out-of-domain [45]. Moreover, the real cost is difficult to extract in most domains [36], and it may result in overfitting to noise and outliers [39].

However, most of the literature has only tried to make the data distribution balanced, few studies have focused on extracting more information from both the majority and minority. Interestingly, under-represent is the main reason for the class imbalance problem mentioned by several researchers [25,45], as a result, new data representations may help the model extract important information about the classification boundary and lead to better performance.

Motivated by the above finding, we aim to design a novel data transformation algorithm to address the imbalanced data problem. The data from different classes will be transformed into a new representation that we can determine the class of a sample based on boundary samples. All information in the majority and minority samples, including even part of the true labels, will be included in the new data representation. Ideally, a more balanced dataset can be generated from our proposed mapping, and this more balanced, new dataset provides a clearer boundary in each different class. A majority voting strategy is utilized to produce the final results based on the model trained on the new dataset. The main contributions of this work are:

1. A new mapping scheme that transforms an imbalanced dataset to a new representation that we can determine the class of a sample based on boundary samples.
2. Samples around the boundary are examined to provide reliable information to enhance the performance of the classifier.
3. Extensive experiments show the different performance improvements on a variety of datasets.

The rest of this paper first gives a brief overview of the recent history of imbalanced learning in Sect. 2, and our methodology will be explained in Sect. 3. Finally, in Sect. 4, the main results of our experiments are discussed and Sect. 5 concludes the paper.

2 Related Work

This section delivers an overview of the technologies utilized to solve the imbalanced learning problem. There are several processes currently applied to alleviate the effects of the class imbalance problem. The most common methods in the literature of imbalanced classification are, such as, sampling methods, cost-sensitive methods, ensemble-based methods [36,39], and data representation methods. A more comprehensive review of various proposed methods in the field can be found in [17]. Imbalanced data is a special case for classification problems where the class distribution is not uniform among the classes. Typically, the instances are composed of two classes: the majority (negative) class and the minority (positive) class. Considering a given dataset S with m samples, $S = \{(x_i, y_i)\}$ where $i = 1, ..., m$. A vector $x_i = (d_1, d_2, ..., d_n)$ can represent a n-dimension sample of the feature space X, and $y_i \in Y$ is the label related to x_i where $Y = \{1...C\}$. The ratio of samples which belongs to majority class S_{maj} and instances that are classified to minority class S_{min} can indicate the difficulty of the class imbalance problem, when the ratio is around 1, the number of samples of different classes are similar, and the performance will not be affected, however, when the ratio is around 0, the difference between classes is obviously, and it has a negative effect on the performance of machine learning algorithms.

Based on the characteristics, the examples of the minority class can be divided into three different groups:

1. Danger: Samples located in the regions around the decision boundary between classes and also the samples located inside overlapping regions of the minority and majority classes.
2. Noise: Isolated pairs or triples of minority class samples, located in the majority class region.
3. Safe: The samples located in the homogeneous regions are populated by examples from the same class.

In some research, noise samples are considered outliers. They are either removed from the samples or considered misclassified samples so their labels are changed to the other class label. Some of the oversampling methods (e.g. MSMOTE [20]) reject noisy spots for the creation of new synthetic samples.

2.1 Sampling Method

In this subsection, We first review some of the sampling approaches. This kind of method attempts to balance the distributions of the original data by oversampling and undersampling.

Sampling methods started from random undersampling approaches, randomly selecting some majority class samples and removing them from the dataset, and random oversampling methods, duplicating some samples of minority class randomly [10]. Although these two kinds of methods are simple to be implemented, there are some defects that cannot be ignored. The random undersampling may lose informative instances, as the samples around the boundary of different classes may be removed from the original dataset. As the same samples repeat several times, the random oversampling may get a specific model which obtains a high accuracy on the training dataset but far worse performance when classifying the unknown instances [18].

Several methods have been proposed to enhance the performance of the undersampling methods and oversampling approaches. Compared to undersampling, the research community paid more attention to improving the oversampling methods, and a classical method is named synthetic minority oversampling technique, (SMOTE) [5] which simulates minority class samples in a space of existing samples. K-nearest neighbours will be found for each minority sample, a simulated data is selected on the line between each data sample and one of its nearest neighbours in the same class. Based on the k-nearest neighbors, MWSMOTE [3] eliminates the noise samples, and then for each sample, it detects the samples that belong to other classes within its k-nearest neighbors. In addition, a different number of synthetic samples will be generated for minority samples according to their distances to the majority samples. There are also several other variants based on SMOTE, for example, borderline-SMOTE [16], MSMOTE [20] algorithms.

However, these algorithms will generate synthetic data samples for each original minority sample. As a result, this process may increase the overlapping between classes, i.e., it may generate noise.

2.2 Cost-Sensitive Methods

The cost-sensitive strategy is another kind of approach to solving the imbalanced learning problem. It assigns a higher weight for the error of misclassifying minority samples which may cause a significant effect [21]. A cost matrice is utilized to define the costs of misclassifying samples into incorrect classes, where C_{ij} indicates the cost of categorizing the samples which belong to class i to class j. Abundant theoretical researches indicate that the class imbalance problem can be solved by the cost-sensitive methods [6].

There are three categories of cost-sensitive methods. The first class assigns different weights to training data [23,44], and the discriminator is trained on the appropriate data space. Modifying the objective of the learning process is another kind of cost-sensitive method [8,26] can pay more attention to the hard samples by modifying the weight of error, some small loss will be eliminated by setting a 0 weight. However, it is difficult to tune the 2 hyper-parameters and adapt to the variational data distribution [24], a method based on the gradient density is proposed in [24], and it suggests removing the weight for both small and large gradients which indicate samples are well classified, and outliers respectively [4]. Moreover, cooperating with other algorithms is the last main option [2].

However, there are a number of limitations to cost-sensitive methods. At first, it is difficult to construct appropriate cost matrices. The values of the cost matrices cannot be determined directly from the data. Moreover, expert opinions cannot be available for each problem. Another reason is that the cost-sensitive methods may change the algorithms, therefore, the cost-sensitive methods are difficult to implement in different models [15].

2.3 Ensemble Methods

A large group of recent research is considered ensemble methods. For example, the EasyEnsemble algorithm and the BalanceCascade algorithm [28]. In the EasyEnsemble algorithm, several subsets of the majority instances are sampled respectively, importantly, the size of each subset is the same as the minority class. All of these subsets are combined with the minority samples as novel datasets. A number of the classifiers of an ensemble learning system are trained based on these hybrid datasets. The final result is the label with the maximum number of votes generated by the discriminators of the EasyEnsemble framework. The BalanceCascade algorithm, which is similar to the EasyEnsemble framework, samples from the majority class and trains the model by the subset and whole minority class instances, after the test process, the majority subset samples which are correctly distinguished by the current model will be removed. This process repeats several times until the model is reached.

However, there are several drawbacks. As the classifiers of these frameworks are based on the Adaboost technique [13], the weights of the whole dataset are assigned after the discriminator is trained once, in addition, to overcome the disadvantages of a single model several classifiers are trained in the framework, as a result, the time consumption is significant.

2.4 Data Representation

A novel approach is proposed as changing the representation of the imbalanced data to enhance the performance of machine learning algorithms [11]. Two arbitrary samples are selected from the training dataset and considered simultaneously in the classification architecture. In the binary classification task with positive samples (P) and negative samples (N), there will be 4 different combinations, (P,P), (P,N), (N,P) and (N,N). As the negative class contains more instances, the class imbalance problem will still exist. When constructing the combinations of majority and majority samples, each sample of the negative class connects with other random $|P|$ majority instances, where $|P|$ is the number of positive instances. To classify new samples, several training samples are selected as reference samples to create novel data representation instances with the test data. The results of all these novel instances will be predicted by the trained classifier, the final result is generated by a majority voting mechanism subsequently. Dumpala et al. claim that a better performance can be received if the reference sample set is composed of majority instances [11]. However, their result may be unreliable when the result of the reference samples is wrong.

3 Proposed Framework and Approach

As mentioned before, the main problem of handling an imbalanced dataset is not the imbalanced ratio [35]. When there is a clear separation between different classes, a hyperplane can correctly divide different samples with an arbitrary imbalanced ratio [30]. As a result, the samples around the boundary which can assist to locate the hyperplane should be found at first. Moreover, transforming the sample to a different format

Algorithm 1. Searching informative samples

Input: $S_{maj}, S_{min}, k_1, k_2, k_3$

S_{maj} : the majority samples

S_{min} : the minority samples

k_1 : number of nearest neighbors for eliminating noise

k_2 : number of nearest neighbors for constructing borderline majority set S_{bmaj}

k_3 : number of nearest neighbors for forming informative minority set S_{imin}

Output: S_{bmaj}, S_{imin}

  1: For $x_i \in S_{min}$, find the k_1 nearest neighbor N_{x_i}

  2: Eliminate the noise instance, whose N_{x_i} belongs to the opposite class, and the left minority samples construct a more reliable minority set S_{smin}

  3: Compute the k_2 nearest majority neighbors for $x_i \in S_{smin}$ as $N_{maj}(x_i)$

  4: Aggregate all the selected majority samples $N_{maj}(x_i)$ as the borderline majority set S_{bmaj}

  5: For $x_i \in S_{bmaj}$, compute the k_3 nearest minority neighbors as $N_{min}(x_i)$

  6: The informative minority set S_{imin} consists of all the $N_{min}(x_i)$

  7: **return** S_{bmaj}, S_{imin}

---

by a map function that captures information from different classes should be processed. And a voting strategy based on the informative samples will be utilized to produce the true label of the testing instances.

3.1 Informative Samples Located

The informative samples are the samples around the hyperplane, which contain more information for training the classifier. The detail about the detection of informative samples can be found in Algorithm 1 which is first proposed in [3]. The novel instances should avoid being noise samples that affect the performance of the classifiers, therefore, for each sample x, the k_1 number of nearest neighbors will be calculated based on the Euclidean distance, if all these neighbors belong to the opposite class, this sample will be identified as a noise sample, and the noise samples in all classes will be eliminated at first to ensure the informative samples are reliable. The k_2 nearest neighbors from the majority class will be detected for each left safe minority sample $x_{smin} \in S_{smin}$ and the selected samples are recognized as borderline majority S_{bmaj}. According to S_{bmaj}, all k_3 nearest neighbors in the minority class formed the informative minority set.

3.2 Extracting Information

General classifiers can be easily skewed up by majority samples, the reason is that the minority does not provide sufficient information. To extract more key information from the imbalance dataset, an approach that considers both majority and minority samples simultaneously based on the transform functions is proposed, particularly, part of the true label is informed to the model (Fig. 1).

In the binary classification task, the minority class refers to positive samples (Pos), and the majority class is considered as negative samples (Neg). Two arbitrary samples

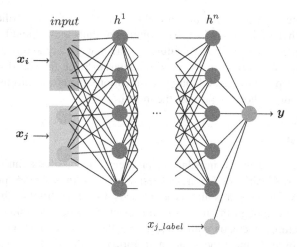

Fig. 1. The structure of the informative S2SL model, where x_i is the first part of the new represent data format, and the x_j is the second part, it should be noted that the true label of x_j is assigned in training and test process.

are selected from the imbalance dataset, as a result, there will be 4 different combinations, (Neg, Neg), (Neg, Pos), (Pos, Neg), and (Pos, Pos). Based on the transformed function, the label will be transformed into a sequence of values to denote the original label of the first sample whether is a positive sample, which can be found below (Table 1).

Table 1. Type of different train method

Transformed label	Data combinations
Negative	(Neg, Neg)
Negative	(Neg, Pos)
Positive	(Pos, Neg)
Positive	(Pos, Pos)

The distribution of the new dataset should be balanced. However, as the number of the majority samples significantly exceed minority samples, the number of (Neg, Neg) is over the other three types after the transformation. To overcome this challenge, the borderline majority and informative minority are utilized in the transformation process. The borderline majority will be chosen as the second part of the (Neg, Neg) transformed samples to achieve an undersampling process, in contrast, the positive samples from the informative minority set are selected to construct the new format data with negative samples. Therefore, the original data will be transform to (P_{imin}, P_{imin}), (P_{imin}, N), (N, P_{imin}), (N, N_{bmaj}).

The neural network is used as the base classifier as it has been certified to be effective in various challenging tasks [40]. The number of units in the input layers is $2 \times d_n$ which is the same as the number of features of the new data representation, where d_n is the number of dimensions in the original dataset. Since the transformed data are either negative or positive (i.e., a binary classification), the cross entropy loss function is employed in our method to optimize the model.

3.3 Model Test

Algorithms that alleviate the imbalanced problem for classifiers can obtain the test results directly by feeding the features of test samples to their models prepared by the training datasets. The reason is that these methods do not modify the dimension of the training data. Even though the map function of our proposed learning method could maintain the feature dimension with the primitive data, the distribution of each feature is different. In order to meet the same input format of the classifier trained by transformed data, the test data are integrated with both the borderline majority samples and informative instances respectively to construct the new data representation, then the test instances are determined by the decompilation of the result of the model.

To enhance the generalization of our method, a voting strategy is applied to producing the final category as most ensemble-based methods do, however, several results of the same test instance are generated by the same classifier. A set of samples are selected from the training dataset for each test instance respectively, and these samples can be used to convert the test data to the target data format which is required by the classifier. It is shown as below:

$$T_{map} = MAP(Test_i, Ref_j) \tag{1}$$

where $Test_i$ is the i_{th} test samples and Ref_j is the j_{th} pre-selected samples. A number of outputs could be produced by the model based on the transformed test dataset. However, the output of the classification model is an intermediate result that cannot present the desired output in the original format. To generate the final result of the new test data, the most class in intermediate predicted results will be selected as the final output of each test data.

The partial order of different samples can be extracted from the training data and presented as the output. To be more specific, the predicted result of the classifier could indicate the test data is more likely to be positive or negative data when compared with the reference data. As a result, the samples around the boundary in the training dataset can lead to a better result. Based on the experiment's result, the combination of borderline majority samples and informative minority samples is recommended to be the reference samples since better performance could be obtained in most situations.

4   Experiments and Results

In this section, we introduce our experiments and analyze the results. Similarly to the other related research, the proposed approach is tested by several real-world imbalanced datasets from different domains, such as bio-informatics and document classification. Their details are presented in Table 2. The number of attributes denotes the features used as the input of the map function. When there is more than one minority class, the datasets are transformed into a binary classification problem by one class against all strategies, which distinguishes the target class samples from all the other classes. The minority listed in the table is the selected class which will be the positive class in the experiments. The imbalance ratio (IR) is another important characteristic that can imply the difficulty of the datasets. It can be defined as:

$$IR = \frac{N_{maj}}{N_{min}} \tag{2}$$

where N_{maj} and N_{min} are the number of instance contained in the majority class and minority class respectively [14]. A slight imbalance will not cause a significant deterioration of the performance of the classifier, in addition, there is not a specific threshold for discriminating whether a dataset is an imbalanced dataset. Normally, a dataset is considered an imbalanced dataset when the IR is more than 1.5 [12]. The datasets we used in this paper are presented in Table 2 and they are ordered by their IR.

Table 2. Characteristics of datasets

Dataset	Attributes	Minority	Examples	IR
Pima	8	268	768	1.90
Glass0	9	70	214	2.01
Vehicle0	18	199	846	3.23
Ecoli1	7	77	336	3.36
Yeast3	8	163	1484	8.11
Pageblock1	10	28	472	15.85
Glass5	9	9	214	22.81
Yeast5	8	44	1484	32.78
Yeast6	8	35	1484	39.15
Abalone	8	32	4172	128.87

All the experiments have been performed on a computer with Xeon (3.7 GHz CPU), 64 GB RAM, and GeForce RTX 2080 Ti for GPU, running python 3.6.2 on Ubuntu 16.04 LTS.

4.1 Results on General Test

To illustrate the performance of our proposed approach, the results will compare with several different state-of-the-art methods, respectively, SMOTE [5], BSMOTE-1 [16], BSMOTE-2 [16], RUSBoot [33], SVMSMOTE [31], MWMOTE [3], S2SL [11], AMOT [37]. In addition, MLP without imbalance strategies is another baseline to be compared, as it is known to have good performance. All the results of different measurement methods are the average value of the 5-fold cross-validation. Normally, for a basic binary classification problem, the minority class is considered the positive class, and the majority class is regarded as the negative class. All the performance measurement values are calculated based on these assumptions.

The F1 scores of all compared algorithms and our proposed method on 10 various datasets are presented in Tables 3, and the best results are highlighted in boldface for each task. Without any enhancement of the imbalanced learning algorithm, the performance of baseline is suffering from the class imbalanced problem, the effect is especially clear when the imbalance rate is really high (e.g. Abalone). The results of our proposed method can significantly enhance the performance, even MLP can produce a pretty good result, for example, Vehicle0. In addition, the proposed algorithm outperforms other state-of-the-art imbalanced learning algorithms in most of the test datasets, the distinct margin can be observed on high imbalanced rate datasets compared to the other methods.

To be more detailed, when compared with the S2SL, the is a 66% (from 0.096 to 0.16) improvement on the rare imbalanced dataset (Abalone), and an obvious increase can also be observed on Glass5 (IR is 22.81) and Yeast6 (IR is 39.15), more than 30% improvement can be obtained by both of them, the performance of Glass5 jumps to 0.591 to 0.808 and the result of Yeast6 climbs to 0.639 from 0.475. Since S2SL achieves a very good result on Vehicle0, the difference between the performance of our approach and S2SL is limited. Except for the extreme scene, our approach can accomplish an 8.79% growth on average when compared with S2SL. The proposed approach provides a significant boost when it comes to another state-of-the-art method, AMOT. More than 10% rise can be found on several rare imbalance datasets, for instance, Abalone and Yeast6. Although our method generates a slight decrease in results on Ecoli1 and Pageblock1, it also provides a competitive performance on the other dataset.

Table 3. Experiments results (F_1)

Method	Pima	Glass0	Vehicle0	Ecoli1	Yeast3	Pageblock1	Glass5	Yeast5	Yeast6	Abalone
MLP	0.554	0.608	0.910	0.635	0.600	0.737	0.600	0.646	0.210	0.000
SMOTE	0.495	0.474	0.376	0.121	0.077	0.091	0.082	0.208	0.058	0.014
Borderlin-1	0.664	0.619	0.894	0.680	0.431	0.677	0.573	0.622	**0.325**	0.067
Borderlin-2	0.498	0.473	0.356	0.110	0.072	0.075	0.082	0.219	0.060	0.013
GSVM	0.600	0.473	0.356	0.110	0.072	0.075	0.082	0.219	0.060	0.013
RUSBoot	0.613	0.637	0.901	0.259	0.502	0.747	0.657	0.649	0.289	0.052
SVMSMOTE	0.677	0.742	0.925	0.776	0.751	0.900	0.613	0.633	0.498	0.054
MWMOTE	0.554	0.608	0.915	0.635	0.660	0.867	0.600	0.646	0.250	0.000
S2SL	0.623	0.686	0.957	0.679	0.721	0.869	0.591	0.709	0.475	0.096
AMOT	0.672	0.753	0.949	**0.826**	0.759	**0.960**	0.666	0.685	0.521	0.068
informative S2S	**0.678**	**0.760**	**0.966**	0.750	**0.781**	0.891	**0.808**	**0.793**	**0.639**	**0.160**

Table 4. Type of different train method

Train method	First Part (Neg, Neg)	Second Part (Neg, Pos)	Third Part (Pos, Neg)	Fourth Part (Pos,Pos)
Normal	(Neg, Sampled Neg)	(Neg, Pos)	(Pos, Neg)	(Pos, Pos)
Borderline Majority	(Neg, BM)	(Neg, Pos)	(Pos, Neg)	(Pos, Pos)
Informative Minority 1	(Neg, Sampled Neg)	(Neg, IM)	(IM, Neg)	(IM, IM)
Informative Minority 2	(Neg, Sampled Neg)	(Neg, Pos)	(Pos, Neg)	(Pos, IM)
Informative Minority 3	(Neg, Sampled Neg)	(Neg, IM)	(Pos, Neg)	(Pos, IM)
AllInfomative	(Neg, BM)	(Neg, IM)	(IM, Neg)	(IM, IM)
AllInfomative 2	(Neg, BM)	(Neg, Pos)	(Pos, Neg)	(Pos, IM)
AllInfomative 3	(Neg, BM)	(Neg, IM)	(Pos, Neg)	(Pos, IM)

Table 5. Results on different reference data (F1 Score)

Train method	Ref data	Pima	Glass0	Vehicle0	Ecoli1	Yeast3	Pageblock1	Glass5	Yeast5	Yeast6	Abalone
Normal	Normal	0.676	0.748	0.956	0.730	0.754	0.851	0.667	0.749	0.487	0.141
	Borderline Majority	0.674	0.748	0.958	0.728	0.775	0.884	0.667	0.770	0.522	**0.160**
	Informative Minority	0.666	0.748	0.954	0.707	0.772	0.865	0.667	0.765	0.526	0.158
	All Informative	0.672	0.748	0.961	0.726	0.776	0.879	0.627	0.772	0.512	0.159
Borderline Majority	Normal	0.647	0.684	0.893	0.607	0.251	0.276	0.505	0.380	0.106	0.020
	Borderline Majority	0.670	0.745	0.955	0.728	0.721	0.872	0.600	0.768	0.490	0.075
	Informative Minority	0.672	0.743	0.957	0.704	0.681	0.735	0.562	0.739	0.494	0.112
	All Infomative	0.673	0.738	**0.966**	0.705	0.775	0.842	**0.808**	0.768	0.548	0.080
Informative Minority 1	Normal	0.634	0.684	0.598	0.645	0.540	0.542	0.480	0.560	0.548	0.075
	Borderline Majority	0.635	0.672	0.653	0.660	0.595	0.784	0.533	0.727	0.572	0.057
	Informative Minority	0.635	0.668	0.633	0.654	0.540	0.674	0.480	0.664	0.583	0.057
	All Infomative	0.633	0.677	0.649	0.666	0.557	0.730	0.480	0.697	0.578	0.057
Informative Minority 2	Normal	0.669	**0.760**	0.953	0.729	0.743	0.846	0.600	0.732	0.546	0.103
	Borderline Majority	0.649	0.756	0.948	0.700	0.751	0.882	0.600	0.762	0.567	0.115
	Informative Minority	0.671	0.756	0.956	0.720	0.775	0.865	0.600	0.770	0.571	0.130
	All Infomative	0.652	0.756	0.965	0.742	0.778	0.878	0.600	0.789	0.552	0.134
Informative Minority 3	Normal	0.675	0.751	0.958	0.735	0.750	0.861	0.667	0.752	0.480	0.105
	Borderline Majority	0.674	0.751	0.959	**0.750**	**0.781**	0.889	0.600	**0.793**	0.535	0.122
	Informative Minority	**0.678**	0.751	0.961	**0.750**	**0.781**	**0.891**	0.667	0.777	0.519	0.097
	All Infomative	0.677	0.751	0.961	**0.750**	**0.781**	0.885	0.600	0.786	0.542	0.105
All Infomative 1	Normala	0.604	0.633	0.509	0.554	0.123	0.301	0.148	0.446	0.191	0.034
	Borderline Majority	0.467	0.600	0.010	0.402	0.247	0.740	0.293	0.315	0.508	0.067
	Informative Minority	0.615	0.678	0.653	0.660	0.603	0.698	0.423	0.676	0.571	0.067
	All Infomative	0.593	0.561	0.548	0.671	0.420	0.747	0.533	0.672	**0.639**	0.067
All Infomative 2	Normal	0.623	0.644	0.520	0.529	0.090	0.200	0.116	0.293	0.075	0.021
	Borderline Majority	0.647	0.533	0.749	0.686	0.581	0.828	0.333	0.703	0.445	0.132
	Informative Minority	0.668	0.706	0.734	0.692	0.627	0.676	0.458	0.717	0.544	0.119
	All Infomative	0.664	0.715	0.854	0.734	0.703	0.827	0.756	0.769	0.552	0.129
All Infomative 3	Normal	0.672	0.745	0.950	0.749	0.742	0.880	0.600	0.757	0.474	0.057
	Borderline Majority	0.669	0.750	0.955	0.731	0.754	0.881	0.600	0.781	0.460	0.117
	Informative Minority	0.671	0.759	0.951	0.740	0.738	0.880	0.600	0.774	0.460	0.073
	All Infomative	0.674	0.747	0.958	0.745	0.754	0.879	0.600	0.777	0.460	0.123

4.2   Results on Different Reference Data

To explain the effects of the different reference data, several classifiers are trained and the test results are enumerated in Table 5. There are 4 different data will be produced during the training process, (Neg, Neg), (Neg, Pos), (Pos, Neg), and (Pos, Pos) as explained in the previous section, the various data employed in different training methods are listed in the Table 4, where the BM is Borderline Majority samples and the IM represents the Informative Minority samples. The normal train method is the same with the S2SL; and the second part of (Neg, Neg) will be replaced by borderline majority samples in the borderline majority training method; there are 3 different training method for utilize the informative minority samples, in the first type, Informative Minority 1 train method, the informative minority samples are used as positive samples in all scenarios, moreover, in the Informative Minority 2 train method, the informative minority samples will only replace the second part of the (Pos, Pos) combinations, in addition, the informative minority samples will be employed as the second parts of (Neg, Pos) and (Pos, Pos) combinations in the Informative Minority 3 train method; as a result, there are 3 different ways to combine the borderline majority samples and the informative minority instance, the borderline majority samples will be used as the reference data in (Neg, Neg) for all 3 different combination train methods, and the informative minority instance are the same with 3 different Informative Minority train methods respectively. Contrastively, there are 4 different reference data types, the normal reference data is recommended by S2SL which can receive a better performance, moreover, the reference data can be sampled in an informative minority dataset, borderline majority dataset, and a combination of both. When the informative minority samples are utilized in all reference scenes, the number of negative samples, for example, (Neg, Neg) and (Neg, Pos) will still be much more than the relevant positive samples, (Pos, Neg) and (Pos, Pos), as a result, the results of Informative Minority and All Informative are worse than the other combinations in almost all scenes with all test reference data situations, however, the limit usage of informative minority samples will produce a relative balance training dataset, therefore, the performance of Informative Minority 2, Informative Minority 3, All Informative 2, All Informative 3 and Border Majority is better. Especially, the best performance can be observed on half of the test datasets by utilizing the Informative Minority 3 training methods, and similar results can be obtained with different reference data, for instance, 0.678 on Pima, and 0.780 on Yeast3. Finally, the test results are also measured using AUC and shown in Table 6.

As the informative minority instances and borderline majority samples are located around the boundary, using both of them can provide more information for classifying the test data nearly the boundary, and a better performance can be produced in most scenarios, which can be demonstrated by most of the best performance are produced by employing the All Informative reference data, for example, 0.966 on Vehicle0 and 0.808 on Glass5. In addition, the informative minority samples can help the classifier to find the true positive data in test data in a strict way, the test results are also good enough in many scenes.

Table 6. Results on different reference data (AUC)

Train method	Ref data	Pima	Glass0	Vehicle0	Ecoli1	Yeast3	Pageblock1	Glass5	Yeast5	Yeast6	Abalone
Normal	Normal	0.750	0.813	0.979	0.840	0.914	0.964	0.845	0.915	0.737	0.589
	Borderline Majority	0.749	0.813	0.978	0.823	0.883	0.948	0.845	0.880	0.806	0.611
	Informative Minority	0.742	0.813	0.978	0.818	0.893	0.964	0.845	0.909	0.832	0.611
	All Infomative	0.747	0.813	0.981	0.825	0.882	0.956	0.843	0.900	0.818	0.611
Borderline Majority	Normal	0.716	0.773	0.961	0.783	0.884	0.825	0.939	0.796	0.745	0.675
	Borderline Majority	0.746	0.812	0.972	0.812	0.847	0.917	0.795	0.852	0.724	0.574
	Informative Minority	0.745	0.813	**0.984**	0.834	**0.954**	0.954	**0.966**	0.910	0.829	0.617
	All Infomative	0.748	0.809	0.984	0.818	0.926	0.943	0.940	0.887	0.820	0.537
Informative Minority	Normal	0.696	0.768	0.777	0.801	0.893	0.911	0.735	0.859	0.859	0.533
	Borderline Majority	0.696	0.757	0.801	0.805	0.905	0.952	0.740	0.898	0.807	0.518
	Informative Minority	0.697	0.753	0.794	0.806	0.819	0.950	0.733	0.910	0.821	0.519
	All Infomative	0.695	0.761	0.803	0.812	0.891	0.956	0.735	0.908	0.820	0.518
Informative Minority 2	Normal	0.743	0.819	0.977	**0.848**	0.913	0.965	0.795	0.914	0.793	0.673
	Borderline Majority	0.730	0.815	0.960	0.796	0.849	0.926	0.795	0.847	0.821	0.625
	Informative Minority	0.744	0.815	0.979	0.829	0.925	0.962	0.795	**0.916**	0.848	0.641
	All Infomative	0.729	0.815	0.979	0.833	0.903	0.955	0.795	0.903	0.847	**0.713**
Informative Minority 3	Normal	0.749	0.816	0.980	0.839	0.936	**0.969**	0.845	**0.916**	0.829	0.673
	Borderline Majority	0.748	0.816	0.982	0.842	0.904	0.949	0.795	0.889	0.819	0.622
	Informative Minority	**0.752**	0.816	0.981	0.842	0.904	0.961	0.845	0.893	0.818	0.551
	All Infomative	0.751	0.816	0.982	0.842	0.904	0.951	0.795	0.895	0.833	0.568
All Infomative	Normal	0.652	0.718	0.702	0.744	0.770	0.828	0.746	0.826	0.823	0.698
	Borderline Majority	0.613	0.497	0.502	0.638	0.580	0.825	0.648	0.611	0.741	0.520
	Informative Minority	0.672	0.766	0.818	0.824	0.897	0.949	0.851	0.905	**0.861**	0.519
	All Infomative	0.665	0.674	0.707	0.797	0.686	0.920	0.833	0.828	0.837	0.519
All Infomative 2	Normal	0.682	0.726	0.706	0.721	0.675	0.737	0.666	0.699	0.689	0.694
	Borderline Majority	0.731	0.679	0.804	0.772	0.738	0.868	0.648	0.794	0.683	0.614
	Informative Minority	0.738	0.790	0.886	0.839	0.950	0.944	0.890	0.915	0.859	0.603
	All Infomative	0.740	0.795	0.924	0.833	0.869	0.935	0.933	0.885	0.807	0.573
All Infomative 3	Normal	0.747	0.808	0.969	0.832	0.859	0.931	0.795	0.863	0.724	0.532
	Borderline Majority	0.744	0.816	0.976	0.819	0.870	0.931	0.795	0.867	0.723	0.552
	Informative Minority	0.746	**0.823**	0.972	0.825	0.871	0.933	0.795	0.861	0.723	0.533
	All Infomative	0.748	0.812	0.978	0.827	0.870	0.934	0.795	0.863	0.723	0.553

5 Conclusion

In this paper, we have introduced a novel approach for solving the imbalanced learning problem. The generation of new samples with different combinations of informative positive instances and border negative samples located around the boundary can help the classifier produce a better performance. By extracting information from the training samples and the instance around the boundary, the classifier can effectively detect the true minority from the test dataset. The effectiveness and performance of the proposed methods are demonstrated by several experiments. The results indicated that the new approaches are much better than before. More research should focus not only on enhancing the stability of the performance, but the generalization around the boundary for each class should also be explored. Especially, classifying the individual instance by utilizing the generated samples.

References

1. Adadi, A.: A survey on data-efficient algorithms in big data era. J. Big Data **8**(1), 1–54 (2021). https://doi.org/10.1186/s40537-021-00419-9
2. Ali, S., Majid, A., Javed, S.G., Sattar, M.: Can-CSC-GBE: developing cost-sensitive classifier with gentleboost ensemble for breast cancer classification using protein amino acids and imbalanced data. Comput. Biol. Med. **73**, 38–46 (2016)
3. Barua, S., Islam, M.M., Yao, X., Murase, K.: Mwmote-majority weighted minority oversampling technique for imbalanced data set learning. IEEE Trans. Knowl. Data Eng. **26**(2), 405–425 (2014)
4. Cao, K., Wei, C., Gaidon, A., Arechiga, N., Ma, T.: Learning imbalanced datasets with label-distribution-aware margin loss. In: Advances in Neural Information Processing Systems, vol. 32 (2019)
5. Chawla, N.V., Bowyer, K.W., Hall, L.O., Kegelmeyer, W.P.: SMOTE: synthetic minority over-sampling technique. J. Artif. Intell. Res. **16**, 321–357 (2002)
6. Chawla, N.V., Japkowicz, N., Kotcz, A.: Special issue on learning from imbalanced data sets. ACM SIGKDD Explor. Newsl. **6**(1), 1–6 (2004)
7. Chen, B., Xia, S., Chen, Z., Wang, B., Wang, G.: RSMOTE: a self-adaptive robust smote for imbalanced problems with label noise. Inf. Sci. **553**, 397–428 (2021)
8. Cheng, F., Zhang, J., Wen, C.: Cost-sensitive large margin distribution machine for classification of imbalanced data. Pattern Recogn. Lett. **80**, 107–112 (2016)
9. Chi, J., et al.: Learning to undersampling for class imbalanced credit risk forecasting. In: 2020 IEEE International Conference on Data Mining (ICDM), pp. 72–81. IEEE (2020)
10. Drummond, C., Holte, R.C., et al.: C4. 5, class imbalance, and cost sensitivity: why under-sampling beats over-sampling. In: Workshop on Learning from Imbalanced Datasets II, vol. 11. Citeseer (2003)
11. Dumpala, S.H., Chakraborty, R., Kopparapu, S.K., Reseach, T.: A novel data representation for effective learning in class imbalanced scenarios. In: IJCAI, pp. 2100–2106 (2018)
12. Fernández, A., LóPez, V., Galar, M., Del Jesus, M.J., Herrera, F.: Analysing the classification of imbalanced data-sets with multiple classes: Binarization techniques and ad-hoc approaches. Knowl.-Based Syst. **42**, 97–110 (2013)
13. Freund, Y., Schapire, R.E., et al.: Experiments with a new boosting algorithm. In: ICML, vol. 96, pp. 148–156. Citeseer (1996)
14. García, V., Sánchez, J.S., Mollineda, R.A.: On the effectiveness of preprocessing methods when dealing with different levels of class imbalance. Knowl.-Based Syst. **25**(1), 13–21 (2012)
15. Haixiang, G., Yijing, L., Shang, J., Mingyun, G., Yuanyue, H., Bing, G.: Learning from class-imbalanced data: review of methods and applications. Expert Syst. Appl. **73**, 220–239 (2017)
16. Han, H., Wang, W.-Y., Mao, B.-H.: Borderline-SMOTE: a new over-sampling method in imbalanced data sets learning. In: Huang, D.-S., Zhang, X.-P., Huang, G.-B. (eds.) ICIC 2005. LNCS, vol. 3644, pp. 878–887. Springer, Heidelberg (2005). https://doi.org/10.1007/11538059_91
17. He, H., Ma, Y.: Imbalanced Learning: Foundations, Algorithms, and Applications. Wiley, Hoboken (2013)
18. Holte, R.C., Acker, L., Porter, B.W., et al.: Concept learning and the problem of small disjuncts. In: IJCAI, vol. 89, pp. 813–818. Citeseer (1989)
19. Hoyos-Osorio, J., Alvarez-Meza, A., Daza-Santacoloma, G., Orozco-Gutierrez, A., Castellanos-Dominguez, G.: Relevant information undersampling to support imbalanced data classification. Neurocomputing **436**, 136–146 (2021)

20. Hu, S., Liang, Y., Ma, L., He, Y.: MSMOTE: improving classification performance when training data is imbalanced. In: Proceedings of the 2009 Second International Workshop on Computer Science and Engineering, vol. 2, pp. 13–17. Citeseer (2009)

21. Johnson, J.M., Khoshgoftaar, T.M.: Survey on deep learning with class imbalance. J. Big Data 6(1), 1–54 (2019). https://doi.org/10.1186/s40537-019-0192-5

22. Krawczyk, B.: Learning from imbalanced data: open challenges and future directions. Progr. Artif. Intell. 5(4), 221–232 (2016). https://doi.org/10.1007/s13748-016-0094-0

23. Lee, J., Sun, Y.G., Sim, I., Kim, S.H., Kim, D.I., Kim, J.Y.: Non-technical loss detection using deep reinforcement learning for feature cost efficiency and imbalanced dataset. IEEE Access 10, 27084–27095 (2022)

24. Li, B., Liu, Y., Wang, X.: Gradient harmonized single-stage detector. In: Proceedings of the AAAI Conference on Artificial Intelligence, vol. 33, pp. 8577–8584 (2019)

25. Liao, T., Taori, R., Raji, I.D., Schmidt, L.: Are we learning yet? A meta review of evaluation failures across machine learning. In: Thirty-Fifth Conference on Neural Information Processing Systems Datasets and Benchmarks Track (Round 2) (2021)

26. Lin, T.Y., Goyal, P., Girshick, R., He, K., Dollár, P.: Focal loss for dense object detection. In: Proceedings of the IEEE International Conference on Computer Vision, pp. 2980–2988 (2017)

27. Liu, W., Wang, L., Chen, J., Zhou, Y., Zheng, R., He, J.: A partial label metric learning algorithm for class imbalanced data. In: Asian Conference on Machine Learning, pp. 1413–1428. PMLR (2021)

28. Liu, X.Y., Wu, J., Zhou, Z.H.: Exploratory undersampling for class-imbalance learning. IEEE Trans. Syst. Man Cybern. Part B (Cybern.) 39(2), 539–550 (2009)

29. Mardani, M., Mateos, G., Giannakis, G.B.: Subspace learning and imputation for streaming big data matrices and tensors. IEEE Trans. Signal Process. 63(10), 2663–2677 (2015)

30. Napierala, K., Stefanowski, J.: Types of minority class examples and their influence on learning classifiers from imbalanced data. J. Intell. Inf. Syst. 46(3), 563–597 (2016). https://doi.org/10.1007/s10844-015-0368-1

31. Nguyen, H.M., Cooper, E.W., Kamei, K.: Borderline over-sampling for imbalanced data classification. In: Proceedings: Fifth International Workshop on Computational Intelligence & Applications, vol. 2009, pp. 24–29. IEEE SMC, Hiroshima Chapter (2009)

32. Qin, H., Zhou, H., Cao, J.: Imbalanced learning algorithm based intelligent abnormal electricity consumption detection. Neurocomputing 402, 112–123 (2020)

33. Seiffert, C., Khoshgoftaar, T.M., Van Hulse, J., Napolitano, A.: RUSboost: a hybrid approach to alleviating class imbalance. IEEE Trans. Syst. Man Cybern.-Part A: Syst. Humans 40(1), 185–197 (2009)

34. Shu, J., et al.: Meta-weight-net: learning an explicit mapping for sample weighting. In: Advances in Neural Information Processing Systems, vol. 32 (2019)

35. Stefanowski, J.: Overlapping, rare examples and class decomposition in learning classifiers from imbalanced data. In: Ramanna, S., Jain, L., Howlett, R. (eds.) Emerging Paradigms in Machine Learning. Smart Innovation, Systems and Technologies, vol. 13, pp. 277–306. Springer, Heidelberg (2013). https://doi.org/10.1007/978-3-642-28699-5_11

36. Tarekegn, A.N., Giacobini, M., Michalak, K.: A review of methods for imbalanced multilabel classification. Pattern Recogn. 118, 107965 (2021)

37. Tripathi, A., Chakraborty, R., Kopparapu, S.K.: A novel adaptive minority oversampling technique for improved classification in data imbalanced scenarios. In: 2020 25th International Conference on Pattern Recognition (ICPR), pp. 10650–10657. IEEE (2021)

38. Vuttipittayamongkol, P., Elyan, E., Petrovski, A.: On the class overlap problem in imbalanced data classification. Knowl.-Based Syst. 212, 106631 (2021)

39. Wang, L., Han, M., Li, X., Zhang, N., Cheng, H.: Review of classification methods on unbalanced data sets. IEEE Access 9, 64606–64628 (2021)

40. Wei, T., Shi, J.X., Li, Y.F., Zhang, M.L.: Prototypical classifier for robust class-imbalanced learning. In: Gama, J., Li, T., Yu, Y., Chen, E., Zheng, Y., Teng, F. (eds.) Advances in Knowledge Discovery and Data Mining. LNCS, vol. 13281, pp. 44–57. Springer, Heidelberg (2022). https://doi.org/10.1007/978-3-031-05936-0_4

41. Wen, G., Wu, K.: Building decision tree for imbalanced classification via deep reinforcement learning. In: Asian Conference on Machine Learning, pp. 1645–1659. PMLR (2021)

42. Xu, Z., Shen, D., Nie, T., Kou, Y., Yin, N., Han, X.: A cluster-based oversampling algorithm combining smote and k-means for imbalanced medical data. Inf. Sci. **572**, 574–589 (2021)

43. Yin, J., Gan, C., Zhao, K., Lin, X., Quan, Z., Wang, Z.J.: A novel model for imbalanced data classification. In: Proceedings of the AAAI Conference on Artificial Intelligence, vol. 34, pp. 6680–6687 (2020)

44. Zhang, C., Gao, W., Song, J., Jiang, J.: An imbalanced data classification algorithm of improved autoencoder neural network. In: 2016 Eighth International Conference on Advanced Computational Intelligence (ICACI), pp. 95–99. IEEE (2016)

45. Zhao, T., Zhang, X., Wang, S.: GraphSMOTE: imbalanced node classification on graphs with graph neural networks. In: Proceedings of the 14th ACM International Conference on Web Search and Data Mining, pp. 833–841 (2021)

Interpretable Decisions Trees via Human-in-the-Loop-Learning

Vladimir Estivill-Castro[1]([☒])[ID], Eugene Gilmore[2], and René Hexel[2][ID]

[1] Universitat Pompeu Fabra, 08018 Barcelona, Spain
vladimir.estivill@upf.edu
[2] Griffith University, Nathan 4111, Australia
{eugene.gilmore,r.hexel}@griffithuni.edu.au

Abstract. Interactive machine learning (IML) enables models that incorporate human expertise because the human collaborates in the building of the learned model. Moreover, the expert driving the learning (human-in-the-loop-learning) can steer the learning objective, not only for accuracy, but perhaps for discrimination or characterisation rules, where isolating one class is the primary objective. Moreover, the interaction enables humans to explore and gain insights into the dataset, and to validate the learned models. This requires transparency and interpretable classifiers. The importance and fundamental relevance of understandable classification has recently been emphasised across numerous applications under the banner of explainable artificial intelligence. We use parallel coordinates to design an IML system that visualises decision trees with interpretable splits beyond plain parallel axis splits. Moreover, we show that discrimination and characterisation rules are also well communicated using parallel coordinates. We confirm the merits of our approach by reporting results from a large usability study.

Keywords: Interactive machine learning ·
Human-in-the-loop-learning · Parallel coordinates · Explainable AI ·
Characteristic rules

1 Introduction

Humans' trust in expert systems has required explanations in human understandable terms [5]. It could be argued that machine learning (ML) was fuelled by the need to transfer human expertise into decision support systems and to reduce the high cost of knowledge engineering. *"It is obvious that the interactive approach to knowledge acquisition cannot keep pace with the burgeoning demand for expert systems; Feigenbaum terms this the 'bottleneck problem'. This perception has stimulated the investigation of ML as a means of explicating knowledge"* [19]. Early reviews on the progress of ML show that the understandability (then named comprehensibility) of the classification was considered vital. *"A definite loss of any communication abilities is contrary to the spirit of AI. AI systems are open to their user who must understand them"* [14]. There is so much to gain by incorporating Human-In-the-Loop-Learning (HILL) in ML

L. A. F. Park et al. (Eds.): AusDM 2022, CCIS 1741, pp. 115–130, 2022.
https://doi.org/10.1007/978-981-19-8746-5_9

tasks, for instance validation or new knowledge elicitation [7, 26]. Today, combining fast heuristic search for classifiers and HILL has received the name of IML [1] because not only are datasets the source of knowledge, but IML also captures the human experience [8]. Supervised learning applications offer extraordinary predictive power but have sacrificed transparency and interpretability of the predictions. Now, new criteria besides predictive accuracy are considered [9,11]; particularly in drought analysis [6], medicine, credit scoring, churn prediction, and bio-informatics [10].

Convolutional Neural Networks (CNNs) are regarded as superior for object classification, face recognition, and automatic understanding of handwriting. Similarly, Support Vector Machines (SVMs) are considered extremely potent for pattern recognition. CNNs, ensembles, and SVMs output "black box" models, since they are difficult to interpret by domain experts [9, 17]. Thus, delivering understandable classification models is an urgent research topic [9, 11]. The most common approach is the production of accurate black-box models followed by extracting explanations [22]. There are two lines of work for delivering explainable models. (1) to build interpretable surrogate models that learn to closely reproduce the output of the black-box model, while regulating aspects such as cluster size for explanation [3]. (2) to produce an explanation for classification of a specific instance [21] or to identify cases belonging to a subset of the feature space where descriptions are suitable [16]. However, there are strong arguments that suggest that real interpretable models must be learnt from the beginning [20].

Learning decision trees from data is one of the pioneer methods that produce understandable models [9,13]. Decision tree learning is now ubiquitous in big data, statistics, and ML. C4.5 (a method based on a recursive approach incorporated into CLS and ID3) is first among the top 10 most used algorithms in data mining. CART (Classification and Regression Trees) is another decision-tree learning method among the top 10 algorithms in data mining. Parallel coordinates (PC) [12] have been used for HILL, but there are few empirical evaluations of its merits. Perhaps the most relevant is an in-depth evaluation of the WEKA [25] package for IML which, however, does not consider the use of PC.

We incorporate PC for exploration of datasets and HILL. We deploy a prototype for the IML of decision-tree classifiers (DTCs). We discuss how this prototype exhibits improvements over many other HILL systems. We emphasise that our prototype enables (1) understanding of learnt classifiers, (2) exploration and insight into datasets, and (3) meaningful exploration by humans. We present how PC can provide a visualisation of specific rules and support the operators' interaction with the dataset to scrutinise specific rules. This enables the construction of characterisation and discrimination rules, which focus on one class above the others. We show that users gain understanding through visualisation by presenting results of a detailed usability study.

In Sect. 2, we review salient HILL systems, where learning classifiers involves dataset visualisations. The advantages of using PC for accuracy have been highlighted elsewhere [15]. But, our review of HILL systems reveals that there is almost no experimental evaluation of the effectiveness of HILL. The largest study is a reproduction with 50 users, while the original WEKA `UserClassifier` paper reported a study with only 5 subjects [25]. Section 3 explains our algorithms and system for HILL. We provide key details of our study that consists of three experiments in Sect. 4. Then, Sect. 5 reports results with over 100 users on our proposed system. We highlight how our system overcomes a number of the shortcomings of the HILL systems we review in Sect. 2.

2   Learning Classifiers Involving Dataset Visualisations

Perhaps the earliest system to profit from the interpretability of decision trees for HILL was PCB's visualisation of an attribute as a coloured bar [2]. This bar is constructed such that each instance of the sorted attribute values is a pixel, coloured corresponding to its class. Users shall visually recognise clusters of a class on an attribute. A DTC is visualised by showing bars with cuts to represent a split on an attribute. Each level of the tree can then be shown as subsets of an attribute bar with splits. A user can participate in learning the tree using this visualisation by specifying where on a bar to split an attribute. The HILL process has some algorithmic support to offer suggestions for splits and to finish subtrees. This visualisation appears particularly effective at showing a large dataset in a way that does not take much screen real-estate. However, the bar representation removes important human domain knowledge; for instance all capability to see actual values of attributes (or the magnitude of difference in values) disappears. This blocks experts from incorporating their knowledge. Moreover, the bar representation restricts classification rules to tests consisting of strictly univariate splits. There is no visualisation of attribute relationships (correlations, inverse correlations, or oblique correlations).

Nested Cavities (NC) [15], is an approach to IML, based on PC [12]. Unlike most other visualisation techniques, PC scale and are not restricted to datasets with a small number of dimensions. Parallel coordinates with 400 dimensions have been used. More attributes are displayed by packing their axes on the side. But decisions being based on over 100 variables are hardly interpretable and understandable. Thus, our prototype uses PC and ML metrics to recommend attributes (and their order) in a visualisation. The operator still can select their preferred number of parallel axes to display. The construction of classifiers with NC is similar to decision-trees, because both approaches follow recursive refinement that results in a decision-tree structure. But, to the best of our knowledge, there are no user-focused evaluations of IML with NC.

When using star-coordinates for dataset visualisation and DTC construction [23,24], each attribute is drawn as an axis on a 2D plane starting from the centre and projected outwards. However, star-coordinates displays suffer similar drawbacks as bar visualisations: users are unable to find subsets of predictive attributes, or ways to discriminate classes. In star-coordinates, the location for visualisation of an instance depends on the value of all attributes, making it impossible to identify boundaries between classes. In contrast, with PC, such separations are readily apparent. With star coordinates, experts cannot explore and interchange attributes with other attributes, even if aware of subsets of predictive attributes. Users can only chose a projection emphasising influential attributes, losing any insight of one attribute's interaction with other attributes. There is no natural interaction with the star-coordinates visualisation where a user can also determine exactly what attribute or attributes are contributing the most to the position of a point in the visualisation. `PaintingClass` [24] extends `StarClass` [23] so the expert can use visualisations of categorical attributes with PC. But the restriction persists for numerical attributes. `PaintingClass` uses evenly distributed categorical values of a categorical attribute along a PC axis. This produces a visualisation with unintended bias. Because `PaintingClass` provides no ML support, and building the classifier is completely human-driven, it could be argued that it is not HILL.

`iVisClassifier` [4] profits from PC, but, to reduce the attributes presented to the user, the dataset is presented after using linear discriminant analysis (LDA). The visualisation uses only the top LDA vectors. But using these new LDA features blocks the user's understanding of the visualisation since each LDA feature is a vector of coefficients over all the original attributes (or dimensions). Heat-maps are displayed in an attempt to provide interpretation of the component features, but they could only possibly have some semantics in the particular application of front-human-portrait face-recognition.

Some empirical evaluation is reported in WEKA's [25] `UserClassifier`. `UserClassifier` is a IML and HILL system for DTCs that shows a scatter plot of only two (user selectable) attributes at a time. A display of small bars for each attribute provides some assistance for attribute relevance. The attribute bar presents the distribution of classes when sorted by that attribute. The user can review the current tree as a node-link diagram in one display, select and expand a node. WEKA's `UserClassifier` is the only one reporting usability studies and it involved only five participants [25]. Later, it was evaluated with 50 university students who had completed 7 weeks of material on ML and DTCs. This study confirmed a number of limitations of the WEKA's `UserClassifier`.

3   Experts Iteratively Construct Decision Trees

We propose a HILL system that addresses the shortcomings of earlier systems, because it uses PC to visualise the training set, the DTC, and also specific rules.

3.1   Using Parallel Coordinates

A PC visualisation draws a vertical axis for each dimension (each attribute). An instance $v = (a_1, a_2, \ldots, a_d)$ (a point in Cartesian coordinates) corresponds to a poly-line in PC that visits value a_n on the axis for attribute A_n (for $n \in \{1, \ldots, d\}$). For HILL, each class has a colour, and the labelled instances of the training set are painted using this colour. Figure 1 shows examples of PC visualisations and the corresponding partially-built decision-tree.

If there are many attributes, a window with a projection onto a subset of the attributes is displayed. On-line Analytical Processing (OLAP) tools enable

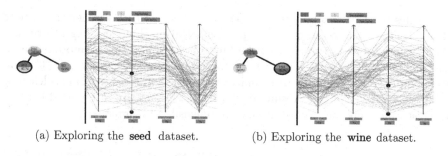

(a) Exploring the **seed** dataset.   (b) Exploring the **wine** dataset.

Fig. 1. Two examples, each exploring a different dataset.

business intelligence practitioners to analyse multidimensional data interactively. Our use of colour concentration allows rapid selection and application of OLAP-type operations on the visible window. For instance, removing one attribute from visualisation and adding another one is a pivot operation. The information gain on an attribute or range within an attribute is used to suggest relevant attributes to the human operator. The user will interactively build a DTC following Hunt's recursive construction: the user picks a leaf-node T to refine the current rule that terminates at T. But, our system provides support for this growth of the tree from leaf T.

1. We colour the corresponding leaf T to illustrate the purity of T (this directly correlates with the classification accuracy of the rule terminating at T). For instance, the left leaf in the tree in Fig. 1a indicates it contains an almost even split of two classes. However, the right leaf is practically pure. The depth of the leaf T inversely correlates with the applicability and generality of that rule terminating at T. Understandability and interpretability also inversely correlate with leaf depth.
2. The system allows the user to select whether to display values of predictability power of attributes, such as the information gain.

All trees are classifiers because the decision at a leaf is a Naïve Bayes decision.

3.2   The Splits the User Shall Apply

A split on one axis alone is commonly a range and this is familiar to DTCs construction as this split belongs to C4.5. However, the PC visualisation allows an *oblique* split that involves two attributes. Thus, the oblique test is interpretable, particularly, because of the point-line duality in PC [12]. For instance, a rule that uses a point between two attributes in PC splits instances into two groups, those that closely follow some linear correlation between two attributes and those that do not. If we use a rectangular region for the split, then we regulate a margin for the above mentioned linear correlation. Figure 2a illustrates the types of splits users can introduce in our system to further a leaf and interactively refine a decision tree. Split (a) and split (b) are familiar from standard decision tree construction. But a rectangular split (c) is an excellent trade-off between interpretability and oblique splits. Besides the interpretability of the splits based on

rectangles between two parallel axes, these splits constitute a richer language to
define DTCs than the standard splits of classical machine learning algorithms.
And although not as powerful as the full oblique splits, this is appropriate, as
full oblique splits are extremely hard to comprehend by humans. DTCs that
use oblique splits are also called multivariate decision trees. Although learning
multivariate decision trees results in shorter trees, they are rarely used because
the test (and thus their rules) are incomprehensible by humans.

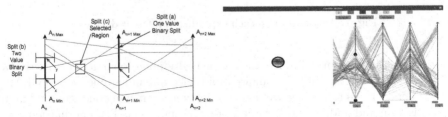

(a) The splits that the user can apply.   (b) A human would chose a better split.

Fig. 2. Visualising decisions as splits.

3.3   Information that Supports Interaction

Our HILL system supports interaction in several ways. For instance, when the
user selects a node in the tree, the visualisation restricts the instances displayed
in the PC-canvas to those that satisfy the splits of the selected node's ancestors.
The user can elect the criteria for ordering attributes in the PC-canvas, which
by default are sorted by a criteria of discriminative power, and among these, the
default is information gain. When the user selects an attribute for the next split,
the system also offers suggestions for the split on the axis, and diverse algorithms
are available (again, information gain, gini-index, etc.). The user can opt to ask
the system to propose a rectangular split. The use can accept or modify the
rectangle suggested by the system. The automatic construction can be restricted
to a node or to a sub-tree. As the user explores proposed sub-trees, the user
gains an understanding of the attributes, and interaction between attributes.
These algorithms support HILL and provide adequate balance between number-
crunching and machine learning support, and user's intervention and interaction
to incorporate human expertise, or for users to discover new insights in the data.

Human pattern spotting has a crucial role on the split selections. Figure 2b
represents a setting where humans easily chose better splits than ML indicators
(such as information gain, gini-index). More purifying are the second attribute
that isolates the brown class and with a robust gap, and the fourth attribute,
which also isolates the green class with also a wide gap. Thus, HILL can deliver
better models. The user can also intervene when ML indicators are offering
similar values, but some attributes are easier to capture or much more readily
available.

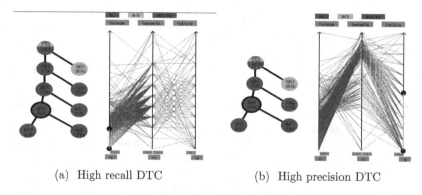

(a) High recall DTC   (b) High precision DTC

Fig. 3. Illustrations of the HILL systems used as a IML to build DTCs with either a high precision or high recall for the black class (coloured red). (Color figure online)

3.4 Visualising the Tree

Figure 1 and Fig. 3 illustrate an interactive visualisation of the tree under construction with the PC-canvas. As we mentioned, the nodes are coloured with a histogram that informs the user about the number of instances of each class that reached the node. This swiftly informs the user of the purity (and thus accuracy) of rules reaching a node. The user can also obtain feedback information about the support (% of the training set instances that reach the node) and confidence (% of correct-class classification) for the rule at the node.

3.5 Visualising Rules

We showed how our HILL system uses PC to effectively assist a user in interactively constructing a DTC. We now demonstrate how we use PC to allow a user to understand a particular classification made by a learnt DTC.

Our approach here supports explorative data mining, where experts are seeking to find characteristic rules or discriminant rules. One of the features of DTCs is their ability to be converted into a decision list. Such a decision list is composed of a series of if-else rules that can be followed to determine the classification of a particular instance. We use a similar idea to visualise the decision path for a single leaf in a DTC. Instead of a textual representation of such a rule however, we can again use PC to graphically represent this decision path. We argue that PC are ideally suited to this task as we can use a series of axes to visualise each component of the rule in the one visualisation. Depending on the depth of the rule and available screen real estate, we may even be able to visualise the entire decision path. Not only this, in our graphical representation, we can visualise the training data and the effect that each component of the decision path has on the resulting subset of selected data. When used in conjunction with our visualisation of the entire DTC and its accompanying statistics for each node, we argue that this gives a human user a profound intuitive and interpretable explanation of a DTC's classification.

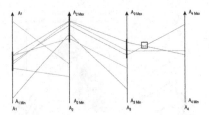
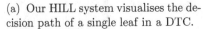
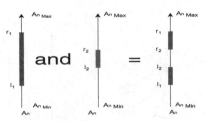

(a) Our HILL system visualises the decision path of a single leaf in a DTC.

(b) Disconnected sections on a single axis by condensing splits that include negation.

Fig. 4. Visualising rules.

Figure 4a shows an example of how we use PC to visualise the decision path to a leaf node in a DTC. Here the leftmost axis is used to represent the split of the root node in the DTC, which will always be the first component in any decision path. In this example, the split for the root node of the decision tree is a simple, single-attribute split on attribute A_1 and is visualised as such with the highlighted range. Between the first and second axes, the start of poly-lines for every instance in the training set is shown. From the second axis, only instances in the training set continue on that matched the split from the previous axis. This allows the user to clearly see what subset of data is selected by each split. The visualisation continues in this manner with poly-lines being terminated once they no longer match a split in the decision path. In this example, the final split in the decision path is a PC region split and is visualised using the last two axes.

Another advantage of visualising a leaf in this way is that users can assess the likelihood that a specific classification is accurate. While performance metrics such as accuracy, ROC, and F1 score capture the performance of the entire tree, it is possible that sections of the model are more (or less) accurate than others. Using this visualisation, users view the amount of the majority class that arrives at a leaf node. When looking at the classification of an instance, users can also see how close to the margins of each split that instance is.

Condensing the Decision Path. In cases where an attribute (or attribute pair in the case of our PC region splits) appears more than once in a decision path, we have the opportunity to reduce the number of attributes required to display the decision path by taking the union of both splits. If using only a single-attribute split, we can create an upper bound on the number of axes required to visualise any decision path equal to the number of attributes in the dataset. When not condensing the decision path in this way, the ordering of axes in PC allows the user to see the depth of each rule in the DTC. For this reason, we provide the user with the option of turning this feature on or off in our HILL system.

Visualising Negated Splits. Negated splits are important since half of all path components are negated splits, i.e. we traverse to the other child node that represents not matching the node's split. For single-attribute splits of the form $l \leq A_n$ the negated split becomes $A_n < l$ and can be represented on a PC axis by simply swapping the highlighted versus the not highlighted region. For a single-attribute test with two split points of the form $l \leq A_n \leq r$, our negated split becomes $A_n < l \vee r < A_n$. This again swaps the highlighted region with the region not highlighted. Although the representation of this two-value split may seem complicated (particularly, when we consider condensing a decision path that contains the same attributes multiple times), swapping of highlighted regions is simplified using De Morgan Laws. Consider a decision path that includes two components using attribute A_n. The first component is of the form $l_1 \leq A_n \leq r_1$ and the second is the negated form of the split $l_2 \leq A_n \leq r_2$, i.e. $A_n < l_2 \vee r_2 < A_n$. Suppose we have the situation where $l_1 < l_2 \wedge r_1 > r_2$. In this case our condensed split becomes $l_1 \leq A_n \leq l_2 \vee r_2 \leq A_n \leq r_1$. To represent this condensed split on a PC axis, we now need to highlight multiple, disconnected sections of the axis. Figure 4b illustrates this situation. Here the green selection on the first axis and the negation of a selection (illustrated as the red region on the second axis) results in two selected ranges on the third axis. In a case where the same attribute is used many times in a decision path, the resulting axis consists of several highlighted disconnected sections (these are disjunctions of intervals). Interestingly, this effect is only possible when using tests containing two split points. We argue that in a HILL system it is only natural that a user will want to use these splits containing two split points to isolate certain sections of data and as such, our system supports visualisation of such condensed rules.

4   Design of the Usability Evaluation

We now report on the evaluation of our prototype by using the online surveying platform Prolific [18]. We obtain quantitative data on the effectiveness of our new visualisation techniques from 104 participants who performed timed tasks. Engagement with the survey requires reviewing 3 concepts applicable to HILL: 1) DTC 2) PC and 3) scatter-plot visualisations. This refresher practice ensures we can rank participants' expertise with ML and exclude anyone not fluent with DTC construction. Moreover, we used Prolific's capability to focus on the following demographics correlated with experience in ML.

- **Which of the following best describes the sector you primarily work in?** IT, Science, Technology, Engineering & Mathematics
- **What is your first language?** English
- **Which of these is the highest level of education you have completed?** Undergraduate degree (BA/BSc/other)
- **Do you have computer programming skills?** Yes

Our survey consists of 3 different experiments. The platform offers participants interaction (and animations) as if they were performing tasks in a system for HILL. Not only does the system display videos and allows users to click on images, and experiment, while moving forward and back through explanations,

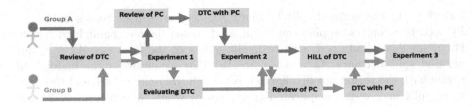

Fig. 5. Composition of survey for participants in Group A and Group B.

but it also enables interactive functionality in all components of building a DTC, including its display, the selection of nodes, and their expansion. In addition, the system allows user configuration of these visualisations.

Experiment 1 examines the effectiveness of visualising a DTC using a node-link diagram with coloured nodes. Experiment 2 evaluates participants' ability to understand the classification of individual instances. This experiment compares the traversal of a DTC against the PC-based visualisation of the path to a leaf. Experiment 3 exhibits video-pairs contrasting two different HILL systems. Showing videos removes the participants' need to gain sufficient expertise with GUI aspects that are not the core of the visualisation. Nevertheless, participants can evaluate how well a system supports users to perform a HILL task.

With probability 1/2, each participant is randomly placed in one of two groups (Group A and Group B). Each group is shown slightly different visualisation when completing tasks in Experiment 1 and Experiment 2 with the aim of contrasting these visualisations' effect on the participants' ability to complete the tasks. Thus, the order of the review material is slightly different. Figure 5 shows the order (work-flow) for each group.

Experiment 1—DTC Node Colouring. In Experiment 1 participants are shown several different DTCs and are required to estimate the accuracy of each DTCas well as select the leaf node that they believe *'is most in need of further refinement to improve the accuracy of the classifier'*. Figure 6 shows the layout for this experiment. The aim is to scrutinise the following hypotheses.

Hypothesis 1. (H 1) The technique of colouring nodes will allow a user to estimate the predictive power of a DTC more accurately.

Hypothesis 2. (H-2) The technique of colouring nodes will allow a user to more quickly estimate the predictive power of a DTC.

Hypothesis 3. (H-3) The technique of colouring nodes will allow a user to more accurately identify the most impure nodes in a DTC.

Hypothesis 4. (H-4) The technique of colouring nodes will allow a user to more quickly identify the most impure nodes in a DTC.

Since accuracy is easier to calculate than the *F*-measure, asking for accuracy ensures that the cognitive load on the participant is reduced. To quantify the

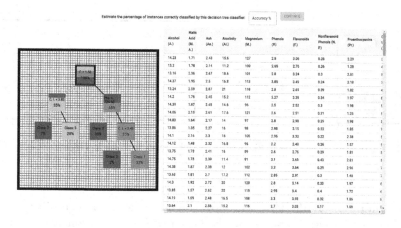

Fig. 6. Layout of survey for Experiment 1.

accuracy of a user's choice for a node in need of improvement, we use a metric $RImp(n)$ describing the impurity of a node in a DTC relative to the most impure node in the tree. Let n be defined as a leaf node in a DTC containing N leaf nodes. Further, let $I(n)$ be defined as the number of instances of training data that reach n whose class is not the majority class of instances reaching n. Our evaluation metric is precisely defined as follows.

Definition 1. We define $I_{max}(N)$ as $I_{max}(N) = \max\{I(n) \mid n \in N\}$.

Definition 2. We define *the relative impurity* $RImp(n)$ of a node by $RImp(n) = (I_{max}(N) - I(n))/I_{max}(N)$.

Each participant is shown 8 different DTCs that were constructed from three datasets available in the UCI-repository (Wine, Cryotherapy, and Seeds). These three datasets exhibit the following relevant properties.

- Reasonably accurate (>90% accuracy) DTC can be learnt for each dataset with small trees sizes that remain interpretable to a human.
- The attributes have humanly understandable names and semantic meaning.

From the datasets, eight DTC were built with a range of different accuracies for two reasons. First, having different accuracies ensures that there are no patterns that participants can use to help them assess the accuracy of any tree. Second, having a range of accuracies ensures that participants' ability to assess the accuracy of a DTC is not dependent on the DTC having a particularly low or high accuracy. For each participant, we record the following information.

- Their prediction of the accuracy of each DTC.
- The time taken to predict the accuracy of each DTC.
- The leaf node selected by the participant as most impure for each DTC.
- The time taken to select the most impure leaf node in each DTC.

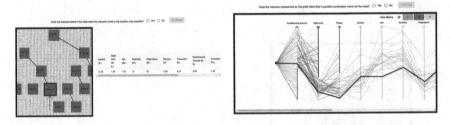

Fig. 7. Layout of Experiment 2: Group A (left) and Group B (right). (Color figure online)

Experiment 2—Rule Visualisation. Figure 7 shows the two different visualisations for Group A and Group B. Group A's visualisation has a DTC on the left and a table on the right. The DTC is a simple node-link diagram with the split criteria for each internal node represented textually. The table contains a single row with the attribute values of one instance. Group A is required to traverse the DTC for the instance shown in the table and determine whether it arrives at a leaf selected by the green arrow.

Group B is shown a PC visualisation of the path to a leaf in a DTC. This PC visualisation shows the entire dataset as well as ranges on several of the PC axes to represent each of the univariate splits on the path to the leaf. The polyline for one instance is shown as a distinct, thick black line. Group B is required to determine if this instance arrives at the leaf being visualised using PC.

Both groups use the same set of instances and DTC. The DTCs used are the same as those in Experiment 1. For each DTC, participants must evaluate three instances. Participants answer whether they think this instance reaches the selected leaf node. This experiment tests the following hypotheses.

Hypothesis 5. (H-5) Participants visualising the path to a node using PC will more accurately determine whether an instance reaches a particular leaf node.

Hypothesis 6. (H-6) Participants visualising the path to a node using PC will more quickly determine whether an instance reaches a particular leaf node.

Experiment 3—Human-in-the-Loop Video Survey. For each pair of videos, one video illustrates a particular HILL task being performed using a system based on the techniques proposed earlier. The other video shows the same task, but carried out using Weka's `UserClassifier`. After viewing both videos, participants are required to express their preference on a five-point Likert scale (Table 1). The survey system randomly decides which HILL system's video will be the first shown. This order remains the same for all video pairs per participant. In total, five different pairs of videos demonstrate a variety of HILL tasks. After answering the Likert-scale questions for each of the five pairs of videos, the survey system asks participants Q6.

Table 1. Questions for Experiment 3.

ID	Question
Q1	Which system do you believe provides a better method of finding splits to build a decision tree classifier?
Q2	Which system allows you to better navigate and understand the current state of a decision tree classifier as it is being constructed?
Q3	Which system allows you to more easily determine how often a tree will predict the class correct class of an instance?
Q4	Which system allows you to more easily find nodes in a decision tree classifier that need additional splits?
Q5	Which system would provide better assistance to you when constructing a decision tree?
Q6	Based on the videos you've seen which system would you prefer to use to build a decision tree classifier?

5   Results

Experiment 1. Table 2a shows the average results for each group. When estimating the accuracy of a DTC, Group B appears to be less accurate than Group A but makes estimates much faster. Similarly, when selecting the most impure leaf in a DTC, the mean relative impurity of leaves selected by Group B was slightly higher, however, the time to choose this leaf was quicker than Group A.

Since the results are not normally distributed by the Shapiro-Wilk test, we use the Wilcoxon-Mann-Whitney test to check for statistically significant differences (as opposed to a T-Test). Performing this statistical test on the accuracy differences and $Rimp$ results in p-values of 0.196 and 0.311, respectively. As such, we cannot reject the null hypotheses that Group A and Group B perform equally and must reject H-1 and H-3. Performing statistical tests on the accuracy time and the leaf selection time results in p-values of 3.65×10^{-10} and 0.006, respectively. Here the differences in time for both tasks are statistically significant: we can accept H-2 and H-4. Although the experiment shows that colouring nodes assist users to more quickly assess the accuracy of a DTC and find leaves requiring refinement, node colouring appears to have little impact on how accurately subjects perform these tasks.

Experiment 2. Table 2b shows the average result (as a % of correctly answered questions) from Group A and Group B. These average accuracies show clear differences in the performance between the two groups. Using the PC-based visualisation, Group B answered 86.7% of questions correctly. This is in contrast to Group A, who only achieved 77.5%. Group B also determines whether an instance reaches a leaf faster than Group A. On average, Group B only required approximately one-quarter of the time that Group A needs to determine whether

Table 2. Results for the first two experiments.

(a) Results for Experiment 1.

Group	Mean accuracy difference	Mean accuracy time (seconds)	Mean *RImp*	Mean leaf choice time (seconds)
A	17.8%	80.2	0.259	17.6
B	19.0%	49.3	0.282	17.1

(b) Results for Experiment 2.

Group	Mean accuracy	Mean time per leaf (seconds)
A	77.5%	24.0
B	86.7%	6.7

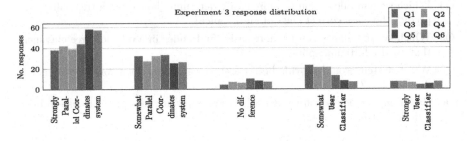

Fig. 8. Distribution of response from Experiment 3.

an instance reaches a leaf. Using the Wilcoxon-Mann-Whitney Test, the average accuracy and time differences are statistically significant with p-values of 1.51×10^{-9} and $< 2.2 \times 10^{-16}$, respectively. As such, H-5 and H-6 are accepted. These results demonstrate a clear advantage to the use of PC to allow humans to interpret DTC. Using PC dramatically decreased the mistakes made when interpreting the splits for a DTC as well as allowing participants to more quickly interpret the series of splits leading to the classification of an instance.

Experiment 3. Figure 8 visualises the distribution of responses received for each question. We can see from these results that there is a clear preference for the PC-based system for all HILL tasks examined. For each of the five pairs of videos, between 66.3% and 79.8% of participants had some preference for the PC-based system. In addition, participant responses from the last question showed 79.8% of participants had an overall preference for the PC-based system.

5.1   Validity Threats

Since participants could not be observed, time measurements may be affected by distractions. Although intended for subjects with limited experience with PC, some may have had previous experience using PC. Finally, all subjects used their monitors, and available screen real-estate may have affected their performance. Regarding external validity, participation was restricted to a bachelor's level, and working in STEM fields. Participants with a higher level of expertise in ML may have found the visualisation techniques more intuitive and performed better.

6   Conclusion

Many have argued around the advantages of HILL for machine learning tasks. Our approach here shows how an effective visualisation can involve a human expert in guiding the construction of interpretable classification models. Moreover, our IML system not only contributes to explainable AI by producing understandable models, it also allows a user to revise the objective of classification accuracy to other important objectives. We have also emphasised that while decision trees rank high as a model for HILL classification, we can strike a suitable balance between multi-variate decision tress and uni-variate decision trees. Users can propose bi-variate splits that remain understandable in our PC-canvas. We have proposed how to visualise and elaborate on characterisation and discrimination rules where users are interested in one class above the others.

References

1. Amershi, S., Cakmak, M., Knox, W.B., Kulesza, T.: Power to the people: the role of humans in interactive machine learning. AI Mag. **35**(4), 105–120 (2014)
2. Ankerst, M., Ester, M., Kriegel, H.P.: Towards an effective cooperation of the user and the computer for classification. In: 6th ACM SIGKDD International Conference Knowledge Discovery and Data Mining, KDD, pp. 179–188. ACM, New York (2000)
3. Blanco-Justicia, A., Domingo-Ferrer, J.: Machine learning explainability through comprehensible decision trees. In: Holzinger, A., Kieseberg, P., Tjoa, A.M., Weippl, E. (eds.) CD-MAKE 2019. LNCS, vol. 11713, pp. 15–26. Springer, Cham (2019). https://doi.org/10.1007/978-3-030-29726-8_2
4. Choo, J., et al.: iVisClassifier: an interactive visual analytics system for classification based on supervised dimension reduction. In: IEEE VAST, pp. 27–34 (2010)
5. Darlington, K.: Aspects of intelligent systems explanation. Univ. J. Control Autom. **1**(2), 40–51 (2013)
6. Dikshit, A., Pradhan, B.: Interpretable and explainable AI (XAI) model for spatial drought prediction. Sci. Total Environ. **801**, 149797 (2021)
7. Estivill-Castro, V.: Collaborative knowledge acquisition with a genetic algorithm. In: 9th ICTAI'97, pp. 270–277. IEEE, California (1997)
8. Fails, J.A., Olsen, D.R.: Interactive machine learning. In: 8th International Conference on Intelligent User Interfaces, IUI '03, pp. 39–45. ACM, New York (2003)
9. Freitas, A.A.: Comprehensible classification models: a position paper. SIGKDD Explor. **15**(1), 1–10 (2013)
10. Freitas, A.A., Wieser, D., Apweiler, R.: On the importance of comprehensible classification models for protein function prediction. IEEE/ACM Trans. Comput. Biology Bioinform. **7**(1), 172–182 (2010)
11. Guidotti, R., et al.: A survey of methods for explaining black box models. ACM CSUR **51**(5), 1–42 (2018)
12. Inselberg, A.: Parallel Coordinates?: Visual Multidimensional Geometry and its Applications. Springer, New York (2009). https://doi.org/10.1007/978-0-387-68628-8
13. Kingsford, C., Salzberg, S.L.: What are decision trees? Nat. Biotechnol. **26**(9), 1011–1013 (2008)

14. Kodratoff, Y.: Chapter 8: Machine learning. In: Knowledge Engineering Volume I Fundamentals, pp. 226–255. McGraw-Hill, USA (1990)

15. Lai, P.L., Liang, Y.J., Inselberg, A.: Geometric divide and conquer classification for high-dimensional data. In: DATA, pp. 79–82. SciTePress (2012)

16. Lakkaraju, H., et al.: Faithful and customizable explanations of black box models. In: AI, Ethics, and Society, AIES 2019, pp. 131–138. ACM, New York (2019)

17. Moore, A., Murdock, V., Cai, Y., Jones, K.: Transparent tree ensembles. In: 41st SIGIR 2018, pp. 1241–1244. ACM, New York (2018)

18. Palan, S., Schitter, C.: Prolific.ac-a subject pool for online experiments. Behav. Exp. Finan. **17**, 22–27 (2018)

19. Quinlan, J.R.: Induction of decision trees. Mach. Learn. **1**(1), 81–106 (1986). https://doi.org/10.1007/BF00116251

20. Rudin, C.: Stop explaining black box machine learning models for high stakes decisions and use interpretable models instead. Nat. Mach. Intell. **1**, 206–215 (2019)

21. Samek, W., et al.: Evaluating the visualization of what a deep neural network has learned. IEEE Trans. Neural Netw. Learn. Syst. **28**(11), 2660–2673 (2017)

22. Samek, W., Müller, K.-R.: Towards explainable artificial intelligence. In: Samek, W., Montavon, G., Vedaldi, A., Hansen, L.K., Müller, K.-R. (eds.) Explainable AI: Interpreting, Explaining and Visualizing Deep Learning. LNCS (LNAI), vol. 11700, pp. 5–22. Springer, Cham (2019). https://doi.org/10.1007/978-3-030-28954-6_1

23. Teoh, S.T., Ma, K.: StarClass: interactive visual classification using star coordinates. In: SIAM International Conference on Data Mining, vol. 112, pp. 178–185 (2003)

24. Teoh, S.T., Ma, K.L.: PaintingClass: interactive construction, visualization and exploration of decision trees. In: 9th SIGKDD'03. ACM, New York (2003)

25. Ware, M., et al.: Interactive machine learning: letting users build classifiers. Int. J. Hum.-Comput. Stud. **55**(3), 281–292 (2001)

26. Webb, G.I.: Integrating machine learning with knowledge acquisition through direct interaction with domain experts. Knowledge-Based Sys. **9**(4), 253–266 (1996)

Application Track

A Comparative Look at the Resilience of Discriminative and Generative Classifiers to Missing Data in Longitudinal Datasets

Sharon Torao Pingi[✉][iD], Md Abul Bashar[iD], and Richi Nayak[iD]

School of Computer Science and Centre for Data Science, Faculty of Science,
Queensland University of Technology, Brisbane, QLD 4000, Australia
`sharon.torao@hdr.qut.edu.au`, {`m1.bashar,r.nayak`}`@qut.edu.au`

Abstract. Longitudinal datasets often suffer from the missing data problem caused by irregular sampling rates and drop-outs, which leads to sub-optimal classification performances. Given the breakthrough of deep generative models in data generation, this paper proposes a conditional Generative Adversarial Network (GAN) based longitudinal classifier (LoGAN) to address this problem in longitudinal datasets. LoGAN is evaluated against commonly used and state-of-art discriminative and generative classifiers. Comparative performance is presented showing the sensitivity of classifiers to data missingness, in both balanced and imbalanced datasets. Results show that the GAN-based models perform on par with other deep learning based models and perform comparatively better on a balanced dataset, while non-deep learning-based discriminative models, in particular the ensemble models, performed better when data was imbalanced. Specifically, F1 scores for LoGAN models were $\geq 80\%$ for up to 20% of missing data rates in the temporal component of the dataset and $\geq 60\%$ for missing rates from 40–100%. Non-deep generative models showed low performance with the introduction of missing data rates.

Keywords: GAN · Missing data · Longitudinal · Disciminative classifiers · Generative classifiers

1 Introduction

Longitudinal datasets are commonly found in many domains such as health and education. They are complex sequential datasets that are multi-dimensional and heterogeneous. They are often collected on a "need basis" which result in subsets of time-point data being missing. This irregularity in sampling causes decreased accuracy in classifiers [15].

Longitudinal datasets have static and dynamic components. Taking the example of the Early School Leaver dataset [8] used in this paper, suppose each

© The Author(s), under exclusive license to Springer Nature Singapore Pte Ltd. 2022
L. A. F. Park et al. (Eds.): AusDM 2022, CCIS 1741, pp. 133–147, 2022.
https://doi.org/10.1007/978-981-19-8746-5_10

student record is an instance. The static component includes variables (or features) that do not change over time, such as the students' date of birth and ethnicity. The dynamic component includes features like students' attendance and academic records collected at defined time points over 11 schooling years. The dynamic component resembles a collection of multivariate time series which contain missing data from irregular sampling and drop-outs. Very few supervised learning methods can cope with longitudinal data [16]. The presence of missing data in such datasets is a complicating factor and makes the selection of an appropriate classifier challenging. Appreciating that each longitudinal dataset has unique temporal and data missingness patterns, having an intuitive understanding of general model behaviour in increased missing data settings can assist in developing or selecting the appropriate classifier as well as in choosing the imputation method and/or model improvement approach.

The deep generative models, particularly GANs [6] that learn the underlying data distributions well through a competitive strategy of adversarially generating and identifying fake samples, can reconstruct observed data for imputation and subsequent classification. Widely used in image and text generation tasks [4,7], GANs have been explored with an auxiliary classification objective for longitudinal datasets with missing data [11,20].

Inspired by the utility of GAN models to condition their learning on additional information, such as labels, to assist auxiliary classification tasks [13], this paper proposes a novel GAN based method, **Lo**ngitudinal **GAN** (LoGAN), to perform classification as a "side" objective to the sample reconstruction task, given partially observed data and class labels. LoGAN attempts better estimates of samples with missing values. We propose three variations of LoGAN to observe how the different GAN models fare against other deep learning and non-deep learning based generative and discriminative classifiers.

Most machine learning methods cannot take in missing values directly. Also, unique temporal patterns in each longitudinal data mean not one imputation method can be considered best for all datasets. Thus, researchers often must consider a classifier and imputation method best suited to the dataset used. Understanding of the behaviour of the different discriminative and generative models (including LoGAN) in the presence of increasing missing temporal data rates will aid researchers decide on a classifier for longitudinal data with irregular sampling and drop-outs. The primary objective of this paper is to present a comparative analysis of the performance of discriminative and generative classifiers and their sensitivity to data missingness, in both balanced and imbalanced datasets.

The following are our main contributions in this paper:

1. We propose a novel longitudinal classifier based on the conditional longitudinal GAN model (LoGAN) with three different implementations of the discriminators and generators based on different deep learning models.
2. We investigate the performance of discriminative and generative models (including LoGAN) against increasing missing time-point data rates that emulate irregular sampling and drop-outs in longitudinal data in both balanced and imbalanced class settings.

2   Background and Related Work

Irregular sampling causes "intermittent missingness" when certain data are intentionally left out for certain instances in given time-points, while drop-outs show missing end-point data. Studies [10] that take into account this informative missingness perform better than those that do not. With this appreciation, we discuss in this study the effects of missing data on classifier performances through emulation of intermittent missingness by irregular sampling and drop-outs.

Research on imputation techniques for longitudinal data shows that the temporal context, patterns and levels of missingness, as well as the choice of classifier affect classification performance [16]. However, not much work has been done in comparing the resilience of discriminative and generative methods for longitudinal classification. The performances of Support Vector Machine (SVM), k-nearest neighbour (kNN), Deep Neural Network (DNN) and Naive Bayes based models on non-longitudinal missing data were compared [1]. Naive Bayes was found to be least influenced by missing data followed by SVM and the kNN classifier was found the most affected. Another study [18] compared the SVMs and Naive Bayes as typical examples of discriminative and generative classifiers for their resilience to missing data. A comparative study [16] on imputation methods on longitudinal datasets was also carried out. However, to the best of our knowledge, this is the first empirical analysis on the effects of increasing missing data rates on a range of discriminative and generative classifiers, alongside a GAN based deep generative classifier on the longitudinal dataset.

There is growing interest in the use of GANs to address irregular sampling for improved longitudinal classification [11,20]. Most of these methods however focus only on the temporal component of the datasets, ignoring static features as being constants throughout the time points. Thus, static features are often not used intentionally to improve temporal sample generation for classification. The proposed LoGAN advances this research area and presents a novel approach to leverage static information and labels as priors to improve an auxiliary classification objective by better estimating irregularly sampled data.

3   LoGAN: A GAN Based Longitudinal Classifier for Missing Data

Discriminative models focus on finding class boundaries within the data space by learning parameters that maximize the posterior $P(Y|X)$ for direct classification. Generative models, on the other hand, first model the data space by learning parameters that maximise the joint probability of $P(X,Y)$, and then estimate $P(Y|X)$ from the Bayes Theorem:

$$P(Y|X) = \frac{P(X|Y)P(Y)}{P(X)} \tag{1}$$

Both classes of supervised models attempt to find $P(Y|X)$ for the most likely class label y, given the observed data $P(X)$. Predicted labels $\hat{y}$ are gained through maximum likelihood estimation with learned parameters θ by:

$$\hat{y} = \underset{y}{\mathrm{argmax}}P(y|X,\theta) \tag{2}$$

For datasets without missing data, discriminative classifiers are preferred over generative classifiers that require an intermediate step [12]. However, given the context, generative models are considered along side discriminative models. Discriminative classifiers are optimised by finding classification rules that return the smallest error rates. Whereas generative models attempt to converge to the best model for $P(X, Y)$, on which GANs have shown better performance among deep generative models [2].

3.1   The LoGAN Approach

The architecture of the proposed LoGAN method is given in Fig. 1. LoGAN is conditioned on static features and class labels to estimate viable longitudinal samples from the sparse temporal data. In this end-to-end implementation, there are two training objectives: (1) the adversarial objective that is trained on the sample generation task; and (2) the classification objective that is trained on the class label prediction task. The joint training of these objectives allows for both to be improved concurrently [13]. As with any GAN, the LoGAN's overall objective in Eq. 3 involves a minimax game of a discriminator model D trying to catch fake samples from a generator model G. The G tries to fool D into believing its samples are real, and the GAN converges when D can no longer tell real samples from highly plausible fakes generated by G.

$$\min_{G} \max_{D_S, D_C} V(D_S, D_C, G) = \mathbb{E}_{x \sim p_r}[\log D_S(X^t|X^s)]$$
$$- \mathbb{E}_{z \sim p_g, x \sim p_r}[\log\left(D_S(G(Z^t|X^s, y))\right)] \tag{3}$$
$$+ \lambda(\mathbb{E}_{(x,y) \sim p_r}[\log D_C(y|(X^t, X^s))]$$

In Eq. 3, D_S and D_C are two output heads of a shared discriminator model D, tasked respectively with Source S discrimination (i.e. determining where the sample is coming from); and Class C discrimination (i.e. predicting the sample's class label (on observed data only)). G is the generator model, and X^s and y are static features and labels, respectively. Input to G is noise sampled from a Gaussian prior $z \sim N(\mu = 0.0, \sigma = 1.0)$, X^s and y. The temporal component of the generated sample $G(z^t|x^s, y)$ is then imputed as $\hat{X}^t = M * X^t - (1 - M) * G(Z^t)$, where M is the random mask applied to the data and $\hat{X}^t$ is passed to D to fool it.

The objective function of LoGAN is modified for G to minimise the chances of D being right about it's samples being fake. This helps prevent D from being over-confident about the fakeness of G's samples too early in training, causing vanishing gradients and convergence failure problems. D_S predicts whether the observed sample is from the training set or drawn from G's estimate (p_g) of the real data distribution p_r. D_S's output is its belief in the realness of the sample its observing, while D_C outputs its belief on the real samples' class labels. The λ parameter regulates D_S and D_C performance during training. Optimal performance is gained when D and G reach a straddle point in trying to outperform each other [6].

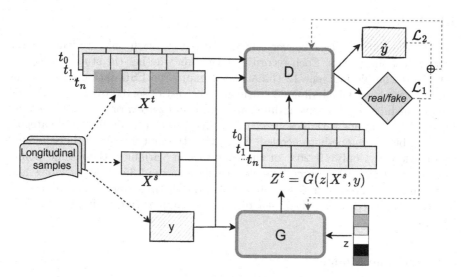

Fig. 1. LoGAN's architecture for longitudinal classification. Black arrows show model inputs and outputs during forward propagation, while red arrows indicate the back-propagation of gradients to update the models. $\mathcal{L}_1$ is and $\mathcal{L}_2$ represent adversarial and classification losses respectively, where D is updated with a linear sum of the two losses. (Color figure online)

Figure 1 describes the LoGAN architecture. D does not see the labels for which G is conditioned on but has to predict. D's final layer is split into output heads D_S and D_C; this means the heads share network weights and can leverage each other's learning [13]. LoGAN is trained on the real data with its inherent missing data patterns, emulated data missingness from drop-outs and irregular sampling are applied through random masks. The model of interest here is the D_C. Once trained, D_C's performance is evaluated on test samples with increasing levels of irregular sampling and drop-out type of missingness.

In this paper, we implement LoGAN with three variations of D and G models; a fully connected network (FCN), a Long-Short Memory Network (LSTM) and a 2-D convolutional neural network (CNN).

4   Experiments

To evaluate the proposed method LoGAN and measure the resilience of selected classifiers to missing data, classifiers are first trained using the training data and then tested on test sets with increasing rates of missing data. The same experiments are performed on balanced and imbalanced versions of the selected dataset. The following subsections introduce the dataset, the experimental setup, the baseline models and evaluation criteria used.

4.1 Dataset

We use the Early School Leavers (ESL) dataset [8] with 35K student records over 11 schooling years. Each record is flagged on whether the student dropped out before completing year 12 (binary outcome). The ESL dataset had an initial 187 variable set including 33 static and the rest observed over time. During pre-processing, unary, derived, duplicate, leaky and zero-variance variables were removed; continuous variables were standardised to a scale of $[0,1]$; and categorical variables were one-hot encoded. Given that the first 6 years of the schooling information had only 10–20% of the time-variant variables, these time points were omitted from the study. The cleaned data set had 35 static variables and 8 variables observed over 5 schooling years (i.e. 5 time steps). The dataset is highly imbalanced with only 7.24% of positive samples (indicating early school leavers).

4.2 Baseline Models

Often, a problem-transformation or algorithm-adaptation approach (or hybrid of these) is taken when classifying longitudinal data. All baselines, except for those based on CNN and LSTM, follow the problem-transformation approach, where all variables are inputted without special consideration to the temporal component of the data. The CNN and LSTM as well as the LoGAN-LSTM and LoGAN-CNN follow the algorithm adaptation approach where static and temporal features are inputted separately.

A total of 15 classifiers are evaluated. Discriminative classifiers include six non-Deep Neural Network (non-DNN) based models, Logistic Regression (LR), SVM, a K-NN classifier (KNN-C), and ensemble models which include Random Forest (RF), XGBoost and a Gradient Boosting Machine (GBM); and three Deep Neural Network (DNN) models based on FCN, LSTM and 2-D CNN. Generative classifiers include three non-deep generative models, namely Naive Bayes, a Gaussian Mixture Model (GMM) and a Quadratic Discriminative Analysis (QDA) model and the three implementation of the deep generative LoGAN model: LoGAN-FCN, LoGAN-LSTM and LoGAN-CNN. With the exception of the DNN and LoGAN models, all models used, including the GridSearchCV algorithm for finding optimal parameter values are from scikit-learn [14] libraries.

4.3 Experimental Setup and Model Design

A 80:20 % train-test split was used with stratified 5-fold cross-validation. Missing values were in-filled with -1 placeholder to represent the value outside of the scaled data range. During training, 10–40% of temporal features were randomly masked to emulate sampling irregularity, and 5–20% of latter time point data masked for drop-out. To test the classifiers' resilience to missing data, temporal features on the test set were randomly masked with 0% to 100% missing rates. Batch sizes of 24 were used on the balanced and imbalanced datasets.

LoGANs' G and D models resembled the design of the DNN models, except for their final layers. The DNN models had a final $Softmax$ layer for the output of class probabilities. LoGAN's G and D shared similar designs but differed in their final layers. The G returned fake time-point data that were imputed as $\hat{X}_t$ before being passed to D. A final $TanH$ layer ensured the data range of generated samples matched the real data. The D had two output layers for D_S and D_C objectives, allowing the tasks to learn correlated information [19]. D's final linear layer is passed through a $Sigmoid$ layer for D_S's real/fake probability output, and through a $Softmax$ layer for D_C's class probabilities.

Like non-DNN models, the FCN based models took in static and temporal features at the same time (along with the label). Both CNN and LSTM models have three initial FCN layers where static features along with label embeddings were fed. The reduced learned features were then passed with temporal features into either the 2x LSTM layers of the LSTM model, or a singled-layered CNN for the CNN model. The outputs to both model types were flattened and passed to a final FCN layer. The 2-D CNN takes in input of dimensions $[timesteps, X^t\_dim]$. Empirically, CNN and LoGAN-CNN required one convolutional layer, no pooling or padding, 3 kernels and stride of 1.

The DNN and LoGAN models were trained on different numbers of epochs to avoid overfitting. A 0% missing rate meant all the temporal data available were present, while 100% missing rate meant none were present; static features and labels helped condition the G. The same experiment was conducted on the original imbalanced data, and repeated on an under-sampled, balanced version of the dataset with ≈5K samples. Optimal parameter values for the LoGAN and DNN models were empirically gained, while GridSearchCV algorithm was used with 10-fold cross-validation for the models. All GAN and DNN models were implemented using python-3.7.4 and Pytorch-1.1.0 while the rest of the models using scikit-learn 0.21.3.

4.4 Evaluation Criteria

The evaluation criteria used include F1-score, Kappa, Area under the ROC curve (AUC), True Positive Rate (TPR) and True Negative Rate (TNR) that determine AUC. Means and standard deviations (stdev) of each metric outcome for the stratified 5-folds and after each epoch and fold for the DNN and LoGAN models are given in Table 1 as $mean(stdev)$. The GAN models were selected based on stability during training, as observed by their discriminator and generator losses.

5 Results and Discussion

In this section, we first discuss the training performance of various machine learning models, and in latter subsections, experimental results for different classifiers on increasing levels of data missingness for both versions of the ESL test dataset (i.e. balanced and imbalanced) are presented.

5.1 Training Performance by Model Setting

All non-DNN discriminative models showed above 94% best accuracy score for the cross validations, except for QDA generative model at 85%. Interestingly, DNN models needed longer training times than the LoGAN models. However, the GANs become highly unstable during training. LoGAN suffered from vanishing gradients where its D became too good in telling fakes early on in training, with insufficient gradients backpropagated to G to improve.

To assist convergence in LoGAN, one-sided label smoothing [17] of [0.7,1.2] was applied in the training of D along with modified loss discussed in Sect. 3.1. Also, batch normalisation (BatchNorm) was used in both the DNN and LoGAN model [17] to improve optimisation. Performance was better when BatchNorm was applied to inputs before being passed to *LeakyReLU*'s activation layers. The ADAM optimiser was used with DNNs converging at a learning rate α equal to 0.0001, 0.001 and 0.003 for FCN, CNN and LSTM respectively, and at α_G and α_D equal to 0.0001, 0.01 and 0.001 for the LoGAN-FCN, LoGAN-CNN and LoGAN-LSTM respectively. The balancing hyper-parameter was set at $\lambda = 0.4$, with $1 - \lambda$ applied on D_S's adversarial loss as it appeared to be improving faster than D_C.

For all non-DNN models, optimal settings by GridSearch on the balanced dataset are: KNN classifier {n_neighbors = 7, weights = distance}, SVM {C = 10, gamma = auto}, RF {criterion = entropy, n_estimators = 500}, QDA {priors = None}, GMM {covariance_type = full, n_components=2, init_params = kmeans, tol = 0.0001}, XGBoost {booster = gbtree, learning_rate = 0.05}. Parameters at default settings are not shown here. Settings for the imbalanced dataset are: SVM {C = 10, gamma = auto}, KNN classifier {n_neighbors = 7, weights = distance}, RF {criterion = entropy, n_estimators = 500}, QDA {priors = None}, GMM {covariance_type = full, n_components=2, init_params = kmeans, tol = 0.0001}, XGBoost {booster = gbtree, learning_rate = 0.05}. LR was set at {C = 0.1} for both datasets. GBM performed best at defaults settings. Naive Bayes had no trainable parameters.

5.2 Performance on Balanced Data

As shown by Table 1 and Fig. 2, the DNN models, especially LSTM, performed better with the balanced dataset. The LoGAN-LSTM started with lower accuracy than the other models, but improved towards higher missing data rates. Given that deep learning algorithms assume an equal distribution of classes,

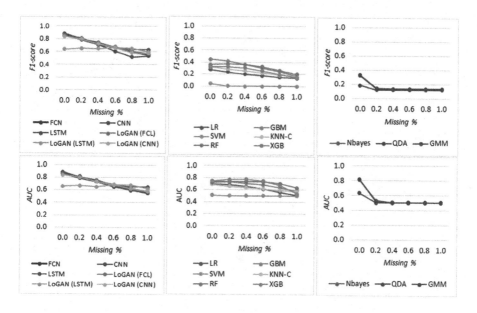

Fig. 2. AUC and F1 scores on balanced ESL dataset

results here are not that surprising. Although, disruptions in the temporal patterns would affect LSTM performance [19].

Results indicate that sufficient contextual information from the static and label data can improve classifier performance for CNN and LSTM, despite sparseness in temporal data. On the flip-side, the models' performance would degrade if insufficient instance-specific information were present to help condition the generated space. Notable performance of the CNN and LoGAN-CNN mean they were able to learn, to a certain degree, the high level features in the data to make good classifications, despite temporal distortions.

Compared to non-DNN models, the DNN and LoGAN models start with higher performance; their Kappa scores indicate strong classification reliability, up to at least 40% data missingness. The reliability of LoGAN classifiers would imply that identifying information within the static features is significantly correlated to the outcome of interest, given that the models were able to learn effectively from the static and label information provided.

The non-DNN discriminative classifiers show <0.5 F1 scores even at 0.0% missing data rate. This indicates their low predictive power and unsuitability in taking in longitudinal data directly. This also shows their lack of flexibility and ability to leverage temporal patterns for learning or considering random effects from the static features separately. Although showing good AUCs, <0.4 Kappa scores early on indicate displayed classification accuracy no better than random guesses. Of the non-DNNs, SVM performs worst. This indicates that sample points were so far dispersed that the SVM model had difficulty separating

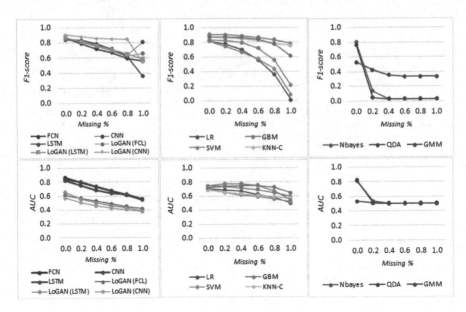

Fig. 3. AUC and F1 scores on imbalanced ESL dataset

them using a linear hyperplane. This is expected given the complex nature of longitudinal datasets.

The non-deep generative classifiers perform poorest, showing lowest TNRs, AUC and F1 scores. At 0% missing rate, the models were able to classify the data with good levels of accuracy, but their performance plunged as soon as missing data is introduced. This indicates that the generative models were not able to learn a joint distribution of the features and associated labels $P(X, y)$ with missing data; being unable to estimate $P(X)$ correctly makes for a biased or weak classifier $P(y|X)$ [3].

5.3   Performance on Imbalanced Data

Table 1 and Fig. 3 show models behaving differently with a skewed class distribution. The non-DNN discriminative models appear the better classifiers on the unbalanced dataset, particularly the ensemble models GBM, RF and XGBoost. While ensemble learning can be effective for imbalanced classification [5], low Kappa scores even at 0% missingness here indicate lower classification strength.

Here, DNNs perform better than the LoGAN models, particularly the LSTM. Similar to the balanced dataset experiments, the CNN and LoGAN-CNN's good performance indicate effective prediction of class labels given their model design, despite missing time step values. Although the FCN shows good F1 scores, its TPR and TNR values indicate this is due to learning majority class very well, but not the minority class. The similar performance is observed for most of the non-DNN classifiers.

The GMM here shows slightly improved F1 scores; however, its Kappa and AUC scores indicate it is still a bad classifier when the missing data is present. The non-deep generative models had high TPRs throughout, being very good at correctly identifying minority class samples, but very low TNR indicate they were also mis-classifying most of the majority class samples as well. This may be due to a high similarity between samples of the different classes.

6   Final Remarks

In summary, the GAN-based models performed better in the balanced dataset, alongside the DNN models; while non-deep classifiers had higher performance in the imbalanced dataset, specially the ensemble models. The worst performing models in both experiments were the traditional (non-deep) generative classifiers, whose performance dropped immediately as soon as missing data was introduced, and stayed consistently low throughout.

As expected, the resilience of the models to increased missing rates would be impacted by the way each classifier learned representations within the data and how they optimised their objective functions. For example, LSTM's sensitivity to temporal patterns and CNN's ability to learn high-level features appeared to give them the edge over other models when trained with most of the temporal data present. Whereas the SVM suffered worst, not being able to linearly separate the samples, and non-deep generative classifiers were not able to learn a density function that explained the data enough to classify it, given missing values.

Interesting insights were gained from the behaviour of the GAN-based models' ability to leverage static features and class labels to improve temporal learning. Static features hold instance-specific information, using these to condition the generator meant it would learn the distribution of the instances in the sample space, so similar instances are grouped together, based on their features as well as class labels. It was found that even with increased missing time-dependent values, sufficient context (in the form of static features and class labels) meant the models were able to train well to give good classification results.

The LoGAN variations were included to investigate the effectiveness of different discriminator and generator models in a GAN setting to address the problem of longitudinal classification with missing data. While showing incredible differences in performance to non-deep generative classifiers and non-DNN models, LoGAN models did not outshine the DNN classifiers. It is argued here that a mechanism, such as an added mean-squared error loss, to assist in the learning of missing patterns would have improved LoGAN's performance to some extent, since without this GANs require fully-observed data during training [9]. However, this was not done in fairness to the other models, which were not aided in learning the missing patterns directly.

It should be noted that the performance of the GAN model depends on the real distribution learned by the discriminator, which serves as the ground truth for the generator to improve on. When there is missing data in the training set, the GAN can struggle to learn the data patterns, particular without some form of assistance, like hints provided to discriminator [20]. This could be another reason for LoGAN not outperforming other models.

The DNN and LoGAN based models showing higher performance with the increasing missing data rates would indicate that these models were benefiting largely from the static and class label information, and would suffer without the information. Correlations in performance observed in DNN and LoGAN models could be due to similarities in the design of LoGANs' generators and discriminators to their DNN model counterparts. Although the use of static features to improve classification was the basis of the LoGAN model designs, they were also included in the DNN models so as not to give LoGAN an unfair advantage. However, it is observed that with this design, the GAN-based models showed strong performance on the balanced dataset.

Many improvements could be made to each class of classifiers for better performance, some suggestions include 1) imbalanced learning with cost sensitive learning, particularly for ensemble models, and further combining ensemble learning with sampling methods, 2) more efficient methods of leveraging random

Table 1. Kappa, True Positive Rate (TPR) and True Negative Rate (TNR) for balanced and imbalanced ESL dataset with increasing missing value rates (Missing %)

Classifier	Experiment 1: Balanced Dataset						Experiment 2: Imbalanced Dataset					
Kappa												
Missing %	0.0	0.2	0.4	0.6	0.8	1.0	0.0	0.2	0.4	0.6	0.8	1.0
LR	0.187(0.010)	0.138(0.006)	0.086(0.007)	0.05(0.012)	0.022(0.007)	0.001(0.000)	0.19(0.009)	0.146(0.005)	0.104(0.008)	0.057(0.006)	0.027(0.003)	0.00(0.00)
GBM	0.395(0.072)	0.362(0.047)	0.295(0.036)	0.229(0.029)	0.161(0.025)	0.087(0.016)	0.400(0.053)	0.376(0.027)	0.314(0.027)	0.249(0.019)	0.162(0.013)	0.085(0.026)
SVM	0.035(0.009)	0.001(0.001)	0.000(0.000)	0.000(0.000)	0.000(0.000)	0.000(0.000)	0.19(0.01)	0.11(0.012)	0.065(0.006)	0.045(0.004)	0.023(0.006)	0.005(0.002)
KNN-C	0.253(0.025)	0.207(0.030)	0.175(0.014)	0.131(0.009)	0.076(0.010)	0.056(0.011)	0.26(0.034)	0.198(0.013)	0.166(0.018)	0.121(0.014)	0.075(0.008)	0.058(0.014)
RF	0.292(0.025)	0.303(0.034)	0.288(0.032)	0.245(0.026)	0.167(0.030)	0.088(0.032)	0.296(0.031)	0.304(0.024)	0.288(0.023)	0.256(0.028)	0.174(0.024)	0.079(0.024)
XGB	0.234(0.007)	0.242(0.010)	0.214(0.006)	0.150(0.004)	0.076(0.005)	0.012(0.001)	0.237(0.006)	0.249(0.006)	0.218(0.007)	0.153(0.006)	0.077(0.004)	0.013(0.003)
FCN	0.332(0.077)	0.278(0.054)	0.204(0.037)	0.134(0.030)	0.071(0.013)	0.019(0.019)	**0.723(0.074)**	**0.583(0.092)**	**0.456(0.096)**	**0.355(0.120)**	**0.224(0.084)**	**0.110(0.163)**
LSTM	**0.764(0.114)**	0.593(0.103)	**0.501(0.154)**	0.344(0.152)	0.301(0.159)	0.291(0.160)	0.300(0.128)	0.240(0.127)	0.148(0.112)	0.096(0.085)	0.065(0.056)	0.023(0.030)
CNN	0.727(0.105)	**0.629(0.104)**	0.464(0.118)	0.299(0.097)	0.192(0.101)	0.108(0.152)	0.291(0.122)	0.249(0.117)	0.178(0.108)	0.117(0.083)	0.065(0.070)	0.050(0.109)
NBayes	0.242(0.009)	0.011(0.001)	0.002(0.000)	0.001(0.000)	0.001(0.000)	0.001(0.000)	0.245(0.003)	0.011(0.00)	0.002(0.00)	0.001(0.00)	0.001(0.00)	0.001(0.00)
QDA	0.229(0.016)	0.005(0.004)	0.002(0.001)	0.001(0.001)	0.001(0.000)	0.000(0.000)	0.224(0.014)	0.003(0.001)	0.004(0.004)	0.00(0.001)	0.00(0.00)	0.001(0.00)
GMM	0.064(0.050)	0.000(0.001)	0.000(0.000)	0.000(0.000)	0.000(0.000)	0.000(0.000)	0.065(0.076)	0.016(0.012)	0.004(0.004)	0.00(0.001)	0.00(0.00)	0.00(0.00)
LoGAN+FCN	0.697(0.012)	0.595(0.014)	0.471(0.017)	**0.365(0.029)**	0.246(0.03)	0.15(0.051)	0.229(0.008)	0.173(0.006)	0.12(0.007)	0.077(0.007)	0.044(0.008)	0.026(0.013)
LoGAN+LSTM	0.319(0.195)	0.336(0.241)	0.295(0.168)	0.354(0.257)	**0.36(0.237)**	0.225(0.025)	0.236(0.082)	0.176(0.015)	0.106(0.025)	0.056(0.018)	0.027(0.012)	0.007(0.042)
LoGAN+CNN	0.67(0.036)	0.592(0.032)	0.453(0.016)	0.362(0.013)	0.225(0.025)	0.156(0.134)	0.298(0.072)	0.206(0.012)	0.119(0.024)	0.065(0.01)	0.03(0.011)	0.01(0.035)
TPR												
Miss %	0.0	0.2	0.4	0.6	0.8	1.0	0.0	0.2	0.4	0.6	0.8	1.0
LR	0.617(0.022)	0.667(0.028)	0.751(0.043)	0.839(0.043)	0.922(0.044)	**1.000(0.000)**	0.626(0.018)	0.662(0.021)	0.749(0.02)	0.82(0.024)	0.919(0.016)	1.00(0.00)
GBM	0.565(0.036)	0.565(0.056)	0.526(0.079)	0.505(0.124)	0.460(0.181)	0.396(0.282)	0.57(0.03)	0.561(0.059)	0.539(0.049)	0.518(0.09)	0.468(0.155)	0.428(0.243)
SVM	0.021(0.005)	0.000(0.001)	0.000(0.000)	0.000(0.000)	0.000(0.000)	0.000(0.000)	0.624(0.019)	0.642(0.019)	0.668(0.023)	0.747(0.025)	0.815(0.024)	0.989(0.011)
KNN-C	0.498(0.025)	0.446(0.019)	0.426(0.019)	0.425(0.031)	0.445(0.065)	0.478(0.129)	0.495(0.018)	0.435(0.021)	0.412(0.034)	0.404(0.033)	0.434(0.059)	0.443(0.124)
RF	0.561(0.021)	0.589(0.021)	0.612(0.026)	0.654(0.050)	0.680(0.072)	0.693(0.128)	0.577(0.012)	0.616(0.011)	0.637(0.02)	0.692(0.03)	0.739(0.049)	0.794(0.078)
XGB	0.697(0.021)	0.768(0.016)	0.836(0.018)	0.891(0.011)	0.935(0.006)	0.967(0.009)	0.695(0.021)	0.774(0.023)	0.831(0.024)	0.888(0.028)	0.929(0.009)	0.959(0.014)
FCN	0.580(0.118)	0.480(0.057)	0.418(0.041)	0.343(0.080)	0.221(0.111)	0.072(0.137)	0.977(0.028)	0.911(0.054)	0.874(0.088)	0.816(0.086)	0.782(0.101)	0.566(0.103)
LSTM	0.937(0.057)	0.918(0.063)	0.895(0.087)	0.822(0.101)	0.808(0.107)	0.834(0.101)	0.861(0.206)	0.733(0.249)	0.657(0.277)	0.657(0.258)	0.762(0.229)	0.884(0.196)
CNN	**0.988(0.027)**	0.987(0.028)	0.990(0.026)	0.989(0.026)	0.986(0.028)	0.747(0.085)	0.898(0.176)	0.858(0.212)	0.779(0.243)	0.742(0.240)	0.720(0.273)	0.307(0.252)
NBayes	0.890(0.026)	**0.999(0.001)**	**1.000(0.000)**	**1.000(0.000)**	**1.000(0.000)**	**1.000(0.000)**	0.898(0.011)	**0.999(0.001)**	**1.00(0.001)**	**1.00(0.00)**	**1.00(0.00)**	**1.00(0.00)**
QDA	0.944(0.068)	0.966(0.052)	0.975(0.044)	0.989(0.016)	**1.000(0.001)**	**1.000(0.000)**	**0.986(0.004)**	**0.999(0.001)**	**1.00(0.00)**	**1.00(0.00)**	**1.00(0.00)**	**1.00(0.00)**
GMM	0.786(0.204)	0.900(0.300)	0.900(0.300)	0.900(0.300)	0.900(0.300)	0.900(0.300)	0.900(0.300)	0.900(0.300)	0.900(0.300)	0.800(0.400)	0.800(0.400)	0.800(0.400)
LoGAN+FCN	0.976(0.016)	0.939(0.028)	0.879(0.033)	0.819(0.033)	0.733(0.032)	0.557(0.088)	0.617(0.024)	0.571(0.024)	0.545(0.032)	0.526(0.029)	0.509(0.056)	0.404(0.049)
LoGAN+LSTM	0.885(0.064)	0.846(0.069)	0.704(0.156)	0.789(0.179)	0.816(0.296)	0.779(0.138)	0.668(0.06)	0.527(0.097)	0.468(0.166)	0.43(0.195)	0.414(0.189)	0.396(0.139)
LoGAN+CNN	0.939(0.034)	0.917(0.02)	0.908(0.008)	0.901(0.013)	0.865(0.021)	0.626(0.134)	0.441(0.114)	0.321(0.018)	0.251(0.046)	0.183(0.022)	0.196(0.02)	0.4(0.157)
TNR												
Missing	0.0	0.2	0.4	0.6	0.8	1.0	0.0	0.2	0.4	0.6	0.8	1.0
LR	0.78(0.003)	0.697(0.016)	0.552(0.038)	0.386(0.062)	0.198(0.07)	0.010(0.001)	0.779(0.002)	0.711(0.009)	0.593(0.017)	0.426(0.022)	0.231(0.024)	0.00(0.00)
GBM	0.921(0.036)	0.909(0.034)	0.889(0.043)	0.858(0.063)	0.821(0.106)	0.768(0.204)	0.923(0.028)	0.918(0.024)	0.896(0.026)	0.867(0.041)	0.824(0.073)	**0.757(0.166)**
SVM	**0.998(0.000)**	**1.000(0.000)**	**1.000(0.000)**	**1.000(0.000)**	**1.000(0.000)**	**1.000(0.000)**	0.78(0.002)	0.667(0.01)	0.556(0.016)	0.44(0.019)	0.301(0.028)	0.043(0.026)
KNN-C	0.877(0.022)	0.867(0.018)	0.853(0.017)	0.815(0.021)	0.732(0.038)	0.675(0.090)	0.882(0.022)	0.867(0.017)	0.853(0.019)	0.815(0.016)	0.738(0.038)	0.703(0.086)
RF	0.879(0.016)	0.876(0.021)	0.859(0.024)	0.815(0.040)	0.724(0.069)	0.574(0.159)	0.796(0.004)	0.869(0.016)	0.852(0.019)	0.809(0.033)	0.708(0.052)	0.494(0.119)
XGB	0.792(0.007)	0.773(0.008)	0.718(0.007)	0.603(0.010)	0.407(0.013)	0.107(0.015)	0.796(0.004)	0.777(0.007)	0.725(0.011)	0.61(0.017)	0.413(0.016)	0.123(0.023)
FCN	0.886(0.061)	0.895(0.032)	0.877(0.021)	0.682(0.179)	0.876(0.076)	0.954(0.098)	0.749(0.062)	0.678(0.072)	0.59(0.090)	0.540(0.094)	0.445(0.114)	0.544(0.093)
LSTM	0.837(0.090)	0.846(0.054)	0.619(0.116)	0.532(0.134)	0.505(0.128)	0.466(0.115)	0.795(0.039)	0.790(0.041)	0.723(0.05)	0.627(0.059)	0.491(0.066)	0.233(0.050)
CNN	0.742(0.093)	0.644(0.105)	0.477(0.106)	0.312(0.093)	0.206(0.095)	0.364(0.102)	0.778(0.044)	0.749(0.059)	0.696(0.079)	0.616(0.117)	0.521(0.120)	**0.789(0.064)**
NBayes	0.729(0.002)	0.070(0.004)	0.011(0.001)	0.010(0.001)	0.010(0.001)	0.010(0.001)	0.728(0.003)	0.071(0.003)	0.011(0.001)	0.01(0.001)	0.01(0.001)	0.01(0.001)
QDA	0.693(0.032)	0.067(0.068)	0.036(0.041)	0.02(0.0180)	0.011(0.002)	0.010(0.001)	0.67(0.018)	0.019(0.004)	0.01(0.001)	0.01(0.001)	0.01(0.001)	0.01(0.001)
GMM	0.480(0.019)	0.104(0.298)	0.100(0.300)	0.100(0.300)	0.100(0.300)	0.100(0.300)	0.536(0.032)	0.291(0.306)	0.22(0.384)	0.203(0.398)	0.200(0.400)	0.200(0.400)
LoGAN+FCN	0.732(0.026)	0.668(0.036)	0.607(0.019)	0.56(0.046)	0.523(0.049)	0.599(0.109)	0.812(0.023)	0.820(0.030)	0.798(0.041)	0.757(0.054)	0.695(0.067)	0.509(0.065)
LoGAN+LSTM	0.443(0.139)	0.499(0.188)	0.599(0.103)	0.57(0.163)	0.548(0.187)	0.465(0.122)	0.757(0.001)	0.76(0.089)	0.493(0.147)	0.282(0.131)	0.145(0.067)	0.428(0.085)
LoGAN+CNN	0.744(0.009)	0.688(0.028)	0.559(0.017)	0.472(0.011)	0.37(0.032)	0.536(0.133)	0.760(0.013)	0.712(0.026)	0.630(0.026)	0.534(0.019)	0.462(0.035)	0.674(0.098)

effects from static features to assist the learning of temporal patterns the dataset, 3) including a reconstructive loss to the GAN's objective function to improve the learning of sample space in the presence of randomly missing data, 4) modifying the GAN architecture to utilise the missing data patterns for classification as a form of input and 5) improving the classification metrics, for instance, setting a threshold for F1 that best suits the data, rather than accepting the default threshold of 0.5.

7   Conclusions

In this paper, we set out to observe the resilience of various discriminative and generative classifiers to increasing missing rates, as compared to the proposed GAN-based longitudinal classifier. Three variations of a conditional GAN model were put forward to observe how different base models of the GAN would fare against other classifiers. The study highlights the utility of GANs in leveraging static features and class labels to improve longitudinal classification, especially when temporal patterns have been disrupted by the presence of missing data.

The study provided an intuitive understanding of how classifier outcomes would be affected given the type and level of data missingness in the longitudinal data. This understanding would assist in the initial choice of classifier and give directions for improvement of outcomes for the models explored. Also, insights into the utility of GANs to take in static and class label priors could be exploited to assist with instance-based sample generation or time-series imputation for longitudinal datasets.

We realise that each model could be improved with further fine tuning of parameters, tweaks in model design or some form of missing values imputation strategy. Future plans include testing with more longitudinal datasets and comparing LoGAN with other state-of-the-art GAN models addressing longitudinal missing data.

References

1. Blomberg, L.C., Hemerich, D., Ruiz, D.D.A.: Evaluating the performance of regression algorithms on datasets with missing data. Int. J. Bus. Intell. Data Min. **8**(2), 105–131 (2013). https://doi.org/10.1504/IJBIDM.2013.057744
2. Bond-Taylor, S., Leach, A., Long, Y., Willcocks, C.G.: Deep generative modelling: a comparative review of VAEs, GANs, normalizing flows, energy-based and autoregressive models. IEEE Trans. Pattern Anal. Mach. Intell. **44**(11), 7327–7347 (2022). https://doi.org/10.1109/TPAMI.2021.3116668
3. Bouchard, G.: Bias-variance tradeoff in hybrid generative-discriminative models. In: Wani, M.A., et al. (eds.) The Sixth International Conference on Machine Learning and Applications, ICMLA 2007, Cincinnati, Ohio, USA, 13-15 December 2007, pp. 124–129. IEEE Computer Society (2007). https://doi.org/10.1109/ICMLA.2007.85
4. Esfahani, S.N., Latifi, S.: Image generation with GANs-based techniques: a survey. Int. J. Comput. Sci. Inf. Technol **11**, 33–50 (2019)

5. Gao, X., et al.: An ensemble imbalanced classification method based on model dynamic selection driven by data partition hybrid sampling. Expert Syst. Appl. **160**, 113660 (2020). https://doi.org/10.1016/j.eswa.2020.113660

6. Goodfellow, I.J., et al.: Generative adversarial nets. In: Ghahramani, Z., Welling, M., Cortes, C., Lawrence, N.D., Weinberger, K.Q. (eds.) Advances in Neural Information Processing Systems 27: Annual Conference on Neural Information Processing Systems, Montreal, Quebec, Canada, pp. 8–13, pp. 2672–2680 (2014). https://proceedings.neurips.cc/paper/2014/hash/5ca3e9b122f61f8f06494c97b1afccf3-Abstract.html

7. Iqbal, T., Qureshi, S.: The survey: text generation models in deep learning. J. King Saud Univ.-Comput. Inf. Sci. (2020)

8. Lamb, S., et al.: Educational opportunity in Australia 2020: who succeeds and who misses out (2020)

9. Li, S.C., Jiang, B., Marlin, B.M.: MisGAN: learning from incomplete data with generative adversarial networks. In: 7th International Conference on Learning Representations, ICLR 2019, New Orleans, LA, USA, 6–9 May 2019. OpenReview.net (2019). https://openreview.net/forum?id=S1lDV3RcKm

10. Ma, Q., et al.: End-to-end incomplete time-series modeling from linear memory of latent variables. IEEE Trans. Cybern. **50**(12), 4908–4920 (2020)

11. Miao, X., Wu, Y., Wang, J., Gao, Y., Mao, X., Yin, J.: Generative semi-supervised learning for multivariate time series imputation. In: Thirty-Fifth AAAI Conference on Artificial Intelligence, AAAI 2021, Thirty-Third Conference on Innovative Applications of Artificial Intelligence, IAAI 2021, The Eleventh Symposium on Educational Advances in Artificial Intelligence, EAAI 2021, Virtual Event, 2–9 February 2021, pp. 8983–8991. AAAI Press (2021). https://ojs.aaai.org/index.php/AAAI/article/view/17086

12. Ng, A.Y., Jordan, M.I.: On discriminative vs. generative classifiers: a comparison of logistic regression and Naive Bayes. In: Dietterich, T.G., Becker, S., Ghahramani, Z. (eds.) Advances in Neural Information Processing Systems 14 [Neural Information Processing Systems: Natural and Synthetic, NIPS 2001(December), pp. 3–8, 2001. Vancouver, British Columbia, Canada], pp. 841–848. MIT Press (2001). https://proceedings.neurips.cc/paper/2001/hash/7b7a53e239400a13bd6be6c91c4f6c4e-Abstract.html

13. Odena, A., Olah, C., Shlens, J.: Conditional image synthesis with auxiliary classifier GANs. In: Precup, D., Teh, Y.W. (eds.) Proceedings of the 34th International Conference on Machine Learning, ICML 2017, Sydney, NSW, Australia, 6–11 August 2017. Proceedings of Machine Learning Research, vol. 70, pp. 2642–2651. PMLR (2017). http://proceedings.mlr.press/v70/odena17a.html

14. Pedregosa, F., et al.: Scikit-learn: machine learning in Python. J. Mach. Learn. Res. **12**, 2825–2830 (2011)

15. Rhemtulla, M., Savalei, V., Little, T.D.: On the asymptotic relative efficiency of planned missingness designs. Psychometrika **81**(1), 60–89 (2016). https://doi.org/10.1007/s11336-014-9422-0

16. Ribeiro, C., Freitas, A.A.: A mini-survey of supervised machine learning approaches for coping with ageing-related longitudinal datasets. In: 3rd Workshop on AI for Aging, Rehabilitation and Independent Assisted Living (ARIAL), held as part of IJCAI-2019, p. 5 (2019)

17. Salimans, T., Goodfellow, I.J., Zaremba, W., Cheung, V., Radford, A., Chen, X.: Improved techniques for training GANs. In: Lee, D.D., Sugiyama,

M., von Luxburg, U., Guyon, I., Garnett, R. (eds.) Advances in Neural Information Processing Systems 29: Annual Conference on Neural Information Processing Systems 2016 (December), pp. 5–10, 2016, Barcelona, Spain, pp. 2226–2234 (2016). https://proceedings.neurips.cc/paper/2016/hash/8a3363abe792db2d8761d6403605aeb7-Abstract.html

18. Shi, H., Liu, Y.: Naïve Bayes vs. Support vector machine: resilience to missing data. In: Deng, H., Miao, D., Lei, J., Wang, F.L. (eds.) AICI 2011. LNCS (LNAI), vol. 7003, pp. 680–687. Springer, Heidelberg (2011). https://doi.org/10.1007/978-3-642-23887-1_86

19. Sun, C., Hong, S., Song, M., Li, H.: A review of deep learning methods for irregularly sampled medical time series data. CoRR abs/2010.12493 (2020). https://arxiv.org/abs/2010.12493

20. Wu, Z., et al.: BRNN-GAN: generative adversarial networks with bi-directional recurrent neural networks for multivariate time series imputation. In: 27th IEEE International Conference on Parallel and Distributed Systems, ICPADS 2021, Beijing, China, 14-16 December 2021, pp. 217–224. IEEE (2021). https://doi.org/10.1109/ICPADS53394.2021.00033

Hierarchical Topic Model Inference by Community Discovery on Word Co-occurrence Networks

Eric Austin[1,2](✉) [iD], Amine Trabelsi[3] [iD], Christine Largeron[4] [iD],
and Osmar R. Zaïane[1,2] [iD]

[1] University of Alberta, Edmonton, AB T6G 2R3, Canada
{eaustin,zaiane}@ualberta.ca
[2] Alberta Machine Intelligence Institute, Edmonton, AB T5J 3B1, Canada
[3] Université de Sherbrooke, Sherbrooke, QC J1K 2R1, Canada
Amine.Trabelsi@USherbrooke.ca
[4] Université Jean Monnet, Saint-Etienne, France
largeron@univ-st-etienne.fr

Abstract. The most popular topic modelling algorithm, Latent Dirichlet Allocation, produces a simple set of topics. However, topics naturally exist in a hierarchy with larger, more general super-topics and smaller, more specific sub-topics. We develop a novel topic modelling algorithm, Community Topic, that mines communities from word co-occurrence networks to produce topics. The fractal structure of networks provides a natural topic hierarchy where sub-topics can be found by iteratively mining the sub-graph formed by a single topic. Similarly, super-topics can by found by mining the network of topic hyper-nodes. We compare the topic hierarchies discovered by Community Topic to those produced by two probabilistic graphical topic models and find that Community Topic uncovers a topic hierarchy with a more coherent structure and a tighter relationship between parent and child topics. Community Topic is able to find this hierarchy more quickly and allows for on-demand sub- and super-topic discovery, facilitating corpus exploration by researchers.

Keywords: Topic modelling · Information networks · Graphs · Natural language processing · Data mining

1 Introduction

Topic modelling discovers the themes of collections of unstructured text documents. Topics can act as features for document classification and indices for information retrieval. However, one of the most important functions of these topics is to assist in the exploration and understanding of large corpora. Researchers in all fields and domains seek to better understand the main ideas and themes of document collections too large for a human to read and summarize. This requires topics that are interpretable and coherent to human users.

Interpretability is a necessary but not sufficient condition for a good topic model. Topics naturally exist in a hierarchy. There are larger, more general super-topics and smaller, more specific sub-topics. "Sports" is a valid topic in that

© The Author(s), under exclusive license to Springer Nature Singapore Pte Ltd. 2022
L. A. F. Park et al. (Eds.): AusDM 2022, CCIS 1741, pp. 148–162, 2022.
https://doi.org/10.1007/978-981-19-8746-5_11

it represents a concept. "Football" and "the Olympics" are also topics. They are not completely distinct from "sports" but rather are sub-topics that fall within sports, i.e. they are child topics of the "Sports" parent topic in the topic hierarchy. Topics also relate to each other to varying degrees. The "movie" topic is more similar to the "television" topic than the "food" topic. This relationship structure is also key to understanding the topical content of a corpus. Topic modelling methods that simply provide the user with a set of topics are not as useful and informative as those that can provide this hierarchy and structure.

Recently, a new domain has emerged where topics can provide utility: conversational agents, which are computer programs that can carry on a human-level conversation. The conversation is an end in itself; the purpose of speaking with a conversational agent is to converse, to be entertained, to express emotion and be supported. The awareness and use of the topics of discussion are key abilities that an agent must possess to be able to carry on a conversation with a human. Previous work has used the detected topic of conversation to enrich the a conversational agent's responses [12]. However, more can be done with topics to improve the abilities of a conversational agent given the right topic model that provides a topic hierarchy and structure. It can be used to detect and control topic drift in the conversation so that the agent's responses make sense in context. If the user is engaged with the current topic, then the agent can stay on topic or detect sub-topics to focus the conversation. The agent can detect super-topics to broaden the range of conversation. The agent should be able to move to related topics or, if the user becomes bored or displeased, jump to dissimilar topics. This type of control over the flow of the conversation is crucial to human communication and is needed for human-computer interaction as well.

The most widely used topic model, Latent Dirichlet Allocation (LDA), only provides a simple set of topics without a hierarchy or structure and has other drawbacks. The number of topics must be specified, requiring multiple runs with different numbers of topics to find the best topics. It performs poorly on short documents. Different runs on the same corpus can produce different topics, especially if the order of the documents is different [25]. Common terms can appear in many different topics, reducing the uniqueness of topics [31].

Neural networks have pushed forward the state-of-the-art in topic modelling. While neural topic models have produced topics of greater coherence, they retain many of the weaknesses of LDA, such as the need to specify the number of topics, while having a tendency to find models with many redundant topics [7] and demanding greater computational resources and specialized hardware.

These drawbacks have inspired us to search for a new approach to topic modelling. We desire a method that can operate quickly on commodity hardware and that provides not only a set of topics but their relationships and a hierarchical structure. It is natural to take an information network-based approach given the growing importance of relational data and graphs in representing complex systems [34]. Our topic modelling algorithm, Community Topic (CT), mines communities from networks constructed from term co-occurrences. These topics are collections of vocabulary terms and are thus interpretable by humans. The fractal nature of the network representation provides a natural topic hierarchy and structure. The topic hyper-vertices form a network with connections of varying strength between the topic vertices derived from the aggregated edges between their constituent word vertices. Super-topics can be mined from this topic network. Each topic itself is also a sub-graph with regions of varying density of connections. This sub-graph can be mined to find sub-topics. Our algorithm has only a single hyperparameter and can run quickly on simple hardware which makes it ideal for researchers from all fields for exploring a document collection.

In this paper, we review related work on topic modelling. We describe our algorithm, how it constructs term co-occurrence networks, and how it mines topics from these networks. We describe how it discovers the topic hierarchy and how this can be done on-the-fly as needed by the user. We empirically evaluate our algorithm and compare it to two probabilistic graphical topic models. Our results show that our approach is able to find a topic hierarchy with a more coherent structure and a tighter relationship between parent and child topics. Community Topic is able to find this hierarchy more quickly and allows for on-demand sub- and super-topic discovery.

2   Related Work

Topic modelling emerged from the field of information retrieval and research to reduce the dimensionality of and more effectively represent documents for indexing, query matching, and document classification. The performance of topic models on these tasks has been surpassed by deep neural models but topic models have become extremely popular tools of applied research both inside and outside of computing science [18]. One early approach is Latent Semantic Analysis (LSA) [10] which decomposes the term-by-document matrix to find vectors representing the latent semantic structure of the corpus and can be viewed as (uninterpretable) topics that relate terms and documents. Another matrix decomposition method is Non-negative Matrix Factorization [23]. Researchers unsatisfied with the lack of a solid statistical foundation to LSA developed Probabilistic Latent Semantic Analysis (pLSA) [17] which posits a generative probabilistic model of the data with the topics as the latent variables.

A drawback of pLSA is that the topic mixture is estimated separately for each document. Latent Dirichlet Allocation (LDA) [5], not to be confused with Linear Discriminant Analysis, was developed to remedy this. LDA is a fully generative model as it places a Dirichlet prior on the latent topic mixture of a document. The probability of a topic z given a document d, $p(z|d; \theta)$, is a multinomial distribution over the topics parameterized by θ where θ is itself a random variable sampled from the prior Dirichlet distribution. The number of topics must be specified and the model provides no topic hierarchy or structure.

There have been many methods developed that attempt to improve upon LDA. Promoting named entities to become the most frequent terms in the document has been tried [22]. In [39], the authors use a process to identify and re-weight words that are topic-indiscriminate. To improve the performance of

LDA on tweets, the authors of [27] pool tweets into longer documents. The Met-aLDA model [41] incorporates meta information such as document labels. The author-topic model [37] extends LDA by conditioning the topic mixture on doc-ument author. The Correlated Topic Model (CTM) [3] models the correlations between topics. The Dynamic Topic Model [4] allows for the modelling of topic evolution over time. Most relevant to our work are two methods that discover a hierarchy of topics. The Hierarchical LDA model (HLDA) [16] models the topic hierarchy using a tree structure. The depth of the tree must be specified but the number of topics is discovered. A flexible generalization of LDA is the Pachinko Allocation Model (PAM) [24]. Like HLDA, PAM allows for a hierarchy of topics but this hierarchy is represented by a directed acyclic graph rather than a tree of fixed depth, allowing for a variety of relationships between topics and terms in the hierarchy, although this structure must be specified by the user.

In recent years, new topic models have emerged based on neural networks. The Embedded Topic Model (ETM) [11] combines word embeddings trained using the Skip-gram algorithm [29] with the LDA probabilistic generative model. Another approach is to use a variational autoencoder (VAE) [20,21] to learn the probability distributions of a generative probabilistic model, as with the neural variational document model (NVDM) [28], the stick-breaking variational autoen-coder (SB-VAE) [30], ProdLDA [36], and Dirichlet-VAE [7]. These models dis-cover topics that are qualitatively different than those found by traditional LDA, although there is debate as to whether they are truly superior [18]. Neural mod-els that provide a topic hierarchy have also been developed. In [40], the authors develop Weibull hybrid autoencoding inference (WHAI) to model multiple lay-ers of priors for deep LDA and thus multiple layers in a topic hierarchy. The number of hyperparameters, complicated training process, and need for special hardware makes this type of model unsuitable for applied researchers seeking a tool for corpus exploration.

3   Community Topic

We call our community detection-based topic modelling algorithm Community Topic (CT). The design of CT was driven by the results of experimentation detailed in our previous paper [2] and the version presented here is the final result of this process of experimentation. The algorithm has three main steps.

3.1   Co-occurrence Network Construction

First, a network is constructed from the document corpus with terms as vertices. An edge exists between a pair of vertices v_i and v_j if the terms t_i and t_j co-occur in the same sentence. The weights of edges are derived from the frequency of co-occurrence. One method is to use the raw count as the edge weight. However, this does not adjust for the frequency of the terms themselves so more common terms will tend to have higher edge weights. An alternative weighting scheme is to use normalized pointwise mutual information (NPMI) between terms (Eq. 1).

$$NPMI(t_i, t_j) = \frac{log \frac{p(t_i, t_j)}{p(t_i)p(t_j)}}{-log(p(t_i, t_j))} \tag{1}$$

NPMI assigns higher values to pairs of terms t_i and t_j whose co-occurrence, $p(t_i, t_j)$, is more frequent than what would be expected if their occurrences in the texts were random, $p(t_i)p(t_j)$. This is normalized to adjust for the frequencies of the terms in the corpus. The edges of the network are thresholded at 0, i.e. those edges with weights ≤ 0 are removed from the network. This is because the community mining algorithm we will use to discover topics uses modularity Q [32] to discover the more densely connected regions of the network. This formula uses the product of the weighted degrees of two vertices to determine the expected value of the strength of their connection if the graph were random, which does not work if a vertex has a negative weighted degree.

$$Q = \frac{1}{2m} \sum_{ij} \left(A_{i,j} - \frac{k_i k_j}{2m} \right) \delta(C_i, C_j) \tag{2}$$

Here m is the sum of weights of all edges in the network, $A_{i,j}$ is the weight of the edge connecting v_i and v_j, k_i (k_j) is the sum of weights of edges incident to v_i (v_j), C_i (C_j) is the assigned community of v_i (v_j), and δ is an indicator function that returns 1 when the two arguments are equal and 0 otherwise.

The distribution of edge weights differs greatly between the raw count and NPMI. The raw count weights follow a power law distribution with the vast majority of edges having very low weight and very few edges with very high weight. This mirrors the power law distribution of term frequencies. Given this distribution of term frequencies, a given edge weight value can carry very different information. An edge weight of 2 could indicate a significant relationship between two terms that occur 5 times each. Between two terms that occur hundreds of times each, an edge weigh of 2 would be noise. When we convert the edge weights to NPMI values, they are scaled to the range $[-1, +1]$ and high values are assigned to edges that represent frequent co-occurrence relative to the frequencies of the connected terms. This distribution resembles a bell curve. We see very few edge weights ≤ 0 that will be removed by thresholding. This indicates that conditioned on co-occurring at least once, two terms are likely to co-occur more often than would be expected by chance. In our experiments we found slightly better results using the NPMI edge weights. We refer the interested reader to our previous work [2] for a visualization of these edge weight distributions.

3.2 Community Mining

Once the co-occurrence network is constructed, CT discovers topics by applying a community detection algorithm. A community is a group of vertices that have a greater density of connections among themselves than they do to vertices outside the group. Many community detection algorithms exist and have been surveyed in other work [9,14,15]. CT employs the Leiden algorithm [38] as this was found to work best in experimentation. The Leiden algorithm has a resolution parameter that is used to set the scale at which communities are discovered. Smaller values of this parameter lead to larger communities being found and larger values lead to smaller communities. This represents the only hyperparameter necessary for CT and is less a value that needs to be carefully tuned for good performance but is rather a way for the user to get communities of a desired size.

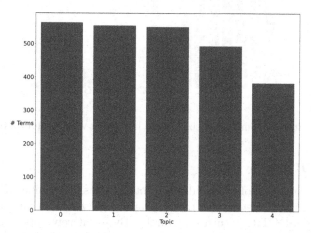

Fig. 1. Distribution of community sizes found by Leiden with resolution parameter 1.0.

Figure 1 Shows the distribution of community sizes found when using a Leiden resolution parameter of 1.0 on the BBC News dataset[1]. CT returns 5 large topics that correspond to the five article categories of the dataset. In Fig. 2, we see the a resolution parameter of 1.5 returns a greater number of small topics with a greater varience of topic size, from hundreds of terms to just a few.

3.3 Topic Filtering and Term Ordering

Once the communities are discovered, small communities of size 2 or less are removed as outliers. Probabilistic graphical topic models such as LDA produce topics that are probability distributions over vocabulary terms. The most important terms for a topic are simply those that have the highest probabilities. The communities discovered by the Leiden algorithm are sets of vertices, so CT needs a way of ranking the terms represented by those vertices. To do so, we take advantage of the graph representation and use internal weighted degree to rank vertices/terms, which is calculated as the sum of weights of edges incident to a vertex that connect to another vertex in the same community/topic. This gives higher values to terms that connect strongly to many terms in the same topic and are thus most representative of that topic. Once the filtering and ordering is complete, the set of topics is returned to the user.

---

[1] https://www.kaggle.com/competitions/learn-ai-bbc/data.

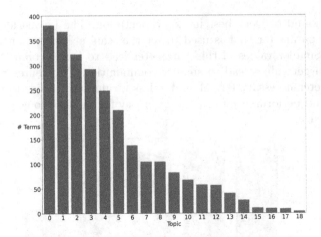

Fig. 2. Distribution of community sizes found by Leiden with resolution parameter 1.5.

3.4 Topic Hierarchy

This basic formulation of CT produces a set of topics like vanilla LDA. However, there exists a natural structure to the graph representation and it is straightforward to adapt CT to return a hierarchy. By iteratively applying community detection to each topic sub-graph, CT discovers the next level of the topic hierarchy. This can be done to a specified depth or we can allow CT to uncover the entire hierarchy by stopping the growth of the topic tree once the produced sub-topics are smaller than three terms. An example of 3 levels of topics discovered on the BBC corpus is show in Fig. 3. The level 1 topics correspond to the 5 article categories of the corpus. Level 2 and then 3 show increasingly specific sub-topics.

The topic hierarchy can also be constructed in a bottom-up fashion. If a low Leiden resolution parameter is initially used, CT produces many small topics. Applying community detection to the network of topic vertices groups these small sub-topics into super-topics. We can see an example of this in Fig. 4 shows the clustering of the initial small topics discovered on the BBC corpus into super-topics which roughly correspond to the 5 article categories of the corpus.

4 Empirical Evaluation

In this section we compare CT to two probabilistic graphical topic models, HLDA and PAM[2]. As the implementation of PAM only allows for two non-root topic layers in the hierarchy we generate a three-level hierarchy for each algorithm for fair comparison, where level 0 is the root topic of all terms in the corpus, level 1 are the super-topics, and level 2 are the sub-topics. PAM requires the number of super- and sub-topics to be specified. We used the number of topics discovered by CT at each level for PAM.

---

[2] https://bab2min.github.io/tomotopy/v0.12.2/en/.

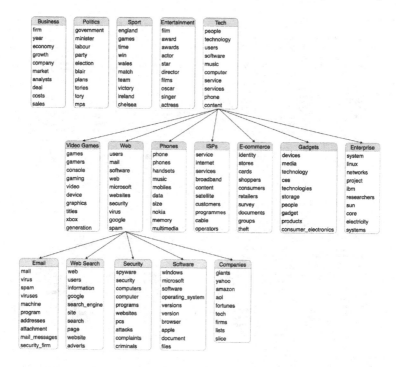

Fig. 3. Hierarchy of BBC corpus topics found by iteratively applying CT algorithm.

4.1 Datasets

We use three datasets to evaluate the different topic modelling approaches: 20Newsgroups[3], Reuters-21578[4], and BBC News[5]. The 20Newsgroups dataset consists of 18,846 posts on the Usenet discussion platform which come from 20 different topics such as "atheism" and "hockey". The Reuters-21578 dataset consists of 21,578 financial articles published on the Reuters newswire in 1987 and have economic and financial topics such as "grain" and "copper". The BBC News dataset consists of 2225 articles in five categories: "business", "entertainment", "politics", "sport", and "tech".

4.2 Preprocessing

We use spaCy[6] to lowercase and tokenize the documents and to identify sentences, parts-of-speech (POS), and named entities. We only detect noun-type entities which are merged into single tokens e.g. the terms "united", "states",

---

[3] https://scikit-learn.org/stable/modules/generated/sklearn.datasets.fetch_20newsgroups.html.

[4] https://huggingface.co/datasets/reuters21578.

[5] https://www.kaggle.com/competitions/learn-ai-bbc/data.

[6] https://spacy.io/.

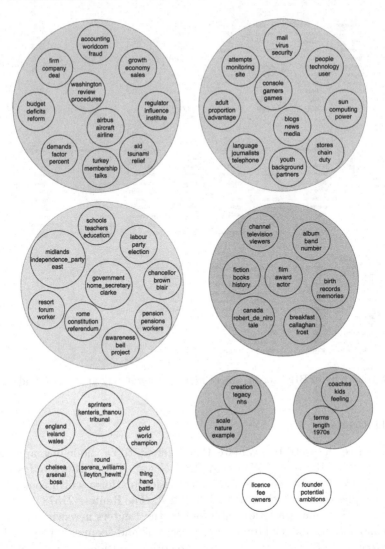

Fig. 4. Super-topics found by applying community detection on network of small topics.

"of", and "america" become "united_states_of_america". While stemming and lemmatization have been commonly used in the topic modelling literature, the authors of [35] found that they do not improve topic quality and hurt model stability so we do not stem or lemmatize. We remove stopwords and terms that occur in >90% of documents. Following [18], we remove terms that appear in fewer than $2(0.02|d|)^{1/log10}$ documents. It was shown in [26] that topic models constructed from noun-only corpora were more coherent so we detect and tag parts-of-speech to be able to filter out non-noun terms as in [8]. This is intuitive as adjectives and verbs can be used in many different contexts, e.g. one can "play the piano", "play baseball", "play the stock market", and "play with someone's heart", but music, sports, finance, and romance are separate topics. Even with nouns there are issues with polysemy, i.e. words with multiple meanings and thus multiple different common contexts. To help with this problem, we use Gensim[7] to extract meaningful n-grams [6]. An n-gram is a combination of n adjacent tokens into a single token so that a term such as "microsoft_windows" can be found and the computer operating system can be distinguished from the windows of a building. We apply two iterations so that longer n-grams such as "law_enforcement_agencies" can be found.

4.3   Evaluation Metrics

To measure the quality of the topics produced by each model, we use two coherence measures: C_V [33] and C_{NPMI} [1]. Both measures have been shown to correlate with human judgements of topic quality with C_V having the strongest correlation [33]. Even though C_V has stronger correlation that C_{NPMI} with human evaluations, C_{NPMI} is more commonly used in the literature [18], possibly due to the extra computation required by C_V. We prefer the C_V measures as, in addition to being more highly correlated with human judgement, it considers the similarity of the contexts of the terms, not just their own co-occurrence. We use Gensim[8] to compute both measures. Each dataset has a train/test split. We train all models on the train documents and evaluate using the test documents. We use the standard 110-term window for C_V and 10-term window for C_{NPMI}. We use the top 5 terms of each topic for evaluation.

To measure the quality of the topic hierarchy, we use two measures proposed in [19]: topic specialization and hierarchical affinity. Topic specialization measures the distance of a topic's probability distribution over terms from the general probability distribution of all terms in the corpus given by their occurrence frequency. We expect topics at higher levels in the hierarchy closer to the root to be more general and less specialized and topics further down the hierarchy to be more specialized. Hierarchical affinity measures the similarity between a super-topic and a set of sub-topics. We expect higher affinity between a parent topic and its children and lower affinity between a parent topic and sub-topics which are not its children.

HLDA produces topics at both levels that are probability distributions over vocabulary terms and are thus compatible with our evaluation metrics without modification. CT produces a list of terms ranked by the internal weighted degree. To calculate specialization and affinity, we convert these to probability distributions by dividing each value by the sum of the values. The super-topics discovered by PAM are distributions over sub-topics. We convert these to distributions over terms by taking the expectation for each term in the sub-topics given the super-topic distribution over sub-topics. Each PAM super-topic distribution gives some non-zero probability to all sub-topics so we need a way

---

[7] https://radimrehurek.com/gensim/.

[8] https://radimrehurek.com/gensim/models/coherencemodel.html.

to distinguish children from non-children. We do this by taking the top 6 most likely sub-topics as the children of a super-topic since we are positing a topic hierarchy with an average of 6 sub-topics per super-topic.

5   Results

Using a Leiden resolution parameter of 1.0, CT finds 5 or 6 super-topics on all datasets and 5, 6, or 7 sub-topics per super topic. We use these average values to guide the PAM model. HLDA finds hundreds of super-topics and about 3 times as many sub-topics. This tendency to find many small topics at all levels leads to poor performance on our metrics and leads to a poor hierarchy where it is common for a child topic to appear in more documents than its parent. PAM performs better, but benefits from using the number of topics discovered by CT.

CT is the fastest of the algorithms, finding the topic hierarchy in under 5 s on all datasets. HLDA takes between 30 s and 5 min while PAM ranges from 10 s to 2 min. All experiments were run on the same laptop with 2.7 GHz dual core processor and 8 GB RAM.

The coherence results are presented in Table 1. We can see that CT achieves the highest coherence scores on all datasets as measured by both metrics except for C_{NPMI} on the 20Newsgroup corpus where PAM comes out on top. PAM achieves the second highest scores in all other cases. HLDA is a distant third with much lower scores. This demonstrates that the topics found by CT will be more interpretable to a human user.

Table 1. Coherence scores for CT, HLDA, PAM on three document corpora. Bold indicates best score for each metric and dataset.

	BBC		20Newsgroups		Reuters	
	C_V	C_{NPMI}	C_V	C_{NPMI}	C_V	C_{NPMI}
CT	**0.641**	**0.079**	**0.645**	0.044	**0.702**	**0.182**
HLDA	0.448	−0.162	0.444	−0.133	0.451	−0.093
PAM	0.600	0.063	0.636	**0.090**	0.555	0.056

Figure 5 shows the specialization scores for each algorithm on the three datasets. We see that both the super-topics (level 1) and the sub-topics (level 2) found by HLDA have a very high specialization. This is consistent with the large number of topics found at both levels but does not match our intuition that topics higher in the hierarchy should be general. PAM produces general topics at level 1 and more specialized topics at level 2, however the super-topics are so general and similar to the overall frequency distribution as to not provide useful information for the user. CT also produces sub-topics that are more specialized than the super-topics. Unlike PAM, the super-topics are themselves specialized and thus useful and informative themselves.

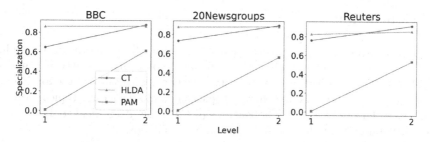

Fig. 5. Topic specialization scores for CT, HLDA, and PAM on three corpora.

Figure 6 shows the hierarchical affinity scores for each algorithm on the three datasets. We see that HLDA has a higher affinity between parent topics and their children than non-children. However, the affinity is very low so the relationship between a super-topic and its sub-topics is very weak. PAM has the opposite problem with high affinities between parent topics and both child and non-child topics. This is because PAM super-topics are distributions over all sub-topics and is consistent with the super-topics being non-specialized. CT parent topics exhibit a high affinity with their children and zero affinity with non-children. This is because the sub-topics are a partition of the super-topic and thus do not overlap with any other super-topic.

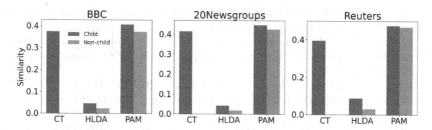

Fig. 6. Hierarchical affinity scores between parent and children and between parent and non-children for CT, HLDA, and PAM on three corpora.

Our experimental results show that CT produces the most coherent and thus interpretable topics and the best topic hierarchy. CT topic hierarchies exhibit higher specialization for sub-topics than super-topics but with enough specialization at both levels to make the topics useful. CT super-topics have a high affinity with their own sub-topics and no affinity with non-child sub-topics. CT is able to produce this coherent topic structure in less time than the other algorithms on commodity hardware.

6   Conclusion

We have presented our novel hierarchical topic modelling algorithm, CT. This method is based on community mining of word co-occurrence networks and is thus fast and takes advantage of the natural network structure. Our experiments show that CT produces more coherent topics and a more cohesive topic hierarchy than either HLDA or PAM. The features of CT make it an ideal tool of corpus exploration and to guide the conversation of a chat bot.

In future, we will extend CT by allowing for overlapping topics. Currently topics are partitions of the vocabulary. A method such as the persona splitting of [13] that creates multiple instances of a vertex would allow for terms to fall into multiple topics. While our method has shown good performance on automated metrics, the real test of a topic model is in its utility for downstream tasks and we plan to integrate CT into a conversational agent to demonstrate its utility.

References

1. Aletras, N., Stevenson, M.: Evaluating topic coherence using distributional semantics. In: Proceedings of the 10th International Conference on Computational Semantics (IWCS 2013)-Long Papers, pp. 13–22 (2013)
2. Austin, E., Zaïane, O., Largeron, C.: Community topic: topic model inference by consecutive word community discovery. In: Proceedings of COLING 2022, the 32nd International Conference on Computational Linguistics (2022)
3. Blei, D., Lafferty, J.: Correlated topic models. In: Advances in Neural Information Processing Systems, vol. 18, p. 147 (2006)
4. Blei, D., Lafferty, J.: Dynamic topic models. In: Proceeding of the 23rd International Conference on Machine Learning, pp. 113–120 (2006). https://doi.org/10.1145/1143844.1143859
5. Blei, D.M., Ng, A.Y., Jordan, M.I.: Latent Dirichlet allocation. J. Mach. Learn. Res. 3(Jan), 993–1022 (2003). https://doi.org/10.1016/B978-0-12-411519-4.00006-9
6. Bouma, G.: Normalized (pointwise) mutual information in collocation extraction. Proc. GSCL 30, 31–40 (2009)
7. Burkhardt, S., Kramer, S.: Decoupling sparsity and smoothness in the Dirichlet variational autoencoder topic model. J. Mach. Learn. Res. 20(131), 1–27 (2019)
8. Chen, J., Zaïane, O.R., Goebel, R.: An unsupervised approach to cluster web search results based on word sense communities. In: 2008 IEEE/WIC/ACM International Conference on Web Intelligence and Intelligent Agent Technology, vol. 1, pp. 725–729. IEEE (2008). https://doi.org/10.1109/WIIAT.2008.24
9. Coscia, M., Giannotti, F., Pedreschi, D.: A classification for community discovery methods in complex networks. Stat. Anal. Data Min. ASA Data Sci. J. 4(5), 512–546 (2011). https://doi.org/10.1002/sam.10133
10. Deerwester, S., Dumais, S.T., Furnas, G.W., Landauer, T.K., Harshman, R.: Indexing by latent semantic analysis. J. Am. Soc. Inf. Sci. 41(6), 391–407 (1990). 10.1002/(sici)1097-4571(199009)41:6<391::aid-asi1>3.0.co;2-9
11. Dieng, A.B., Ruiz, F.J.R., Blei, D.M.: Topic modeling in embedding spaces. Trans. Assoc. Comput. Linguist. 8, 439–453 (2020). https://doi.org/10.1162/tacl_a_00325

12. Dziri, N., Kamalloo, E., Mathewson, K., Zaïane, O.R.: Augmenting neural response generation with context-aware topical attention. In: Proceedings of the First Workshop on NLP for Conversational AI, pp. 18–31 (2019). https://doi.org/10.18653/v1/W19-4103

13. Epasto, A., Lattanzi, S., Paes Leme, R.: Ego-splitting framework: from non-overlapping to overlapping clusters. In: Proceedings of the 23rd ACM SIGKDD International Conference on Knowledge Discovery and Data Mining, pp. 145–154 (2017)

14. Fortunato, S.: Community detection in graphs. Phys. Rep. **486**(3–5), 75–174 (2010). https://doi.org/10.1016/j.physrep.2009.11.002

15. Fortunato, S., Hric, D.: Community detection in networks: a user guide. Phys. Rep. **659**, 1–44 (2016). https://doi.org/10.1016/j.physrep.2016.09.002

16. Griffiths, T., Jordan, M., Tenenbaum, J., Blei, D.: Hierarchical topic models and the nested Chinese restaurant process. In: Advances in Neural Information Processing Systems, vol. 16 (2003)

17. Hofmann, T.: Probabilistic latent semantic indexing. In: Proceedings of the 22nd Annual International ACM SIGIR Conference on Research and Development in Information Retrieval, pp. 50–57 (1999). https://doi.org/10.1145/312624.312649

18. Hoyle, A., Goel, P., Hian-Cheong, A., Peskov, D., Boyd-Graber, J., Resnik, P.: Is automated topic model evaluation broken? The incoherence of coherence. In: Advances in Neural Information Processing Systems, vol. 34 (2021)

19. Kim, J.H., Kim, D., Kim, S., Oh, A.: Modeling topic hierarchies with the recursive Chinese restaurant process. In: Proceedings of the 21st ACM International Conference on Information and Knowledge Management, pp. 783–792 (2012)

20. Kingma, D.P., Welling, M.: Auto-encoding variational bayes. In: Proceedings of the International Conference on Learning Representations (ICLR) (2014)

21. Kingma, D.P., Welling, M., et al.: An introduction to variational autoencoders. Found. Trends Mach. Learn. **12**(4), 307–392 (2019). https://doi.org/10.1561/9781680836233

22. Krasnashchok, K., Jouili, S.: Improving topic quality by promoting named entities in topic modeling. In: Proceedings of the 56th Annual Meeting of the Association for Computational Linguistics (Volume 2: Short Papers), pp. 247–253 (2018). https://doi.org/10.18653/v1/P18-2040

23. Lee, D.D., Seung, H.S.: Learning the parts of objects by non-negative matrix factorization. Nature **401**(6755), 788–791 (1999). https://doi.org/10.1038/44565

24. Li, W., McCallum, A.: Pachinko allocation: DAG-structured mixture models of topic correlations. In: Proceedings of the 23rd International Conference on Machine Learning, ICML 2006, pp. 577–584. Association for Computing Machinery, New York (2006). https://doi.org/10.1145/1143844.1143917

25. Mantyla, M.V., Claes, M., Farooq, U.: Measuring LDA topic stability from clusters of replicated runs. In: Proceedings of the 12th ACM/IEEE International Symposium on Empirical Software Engineering and Measurement, pp. 1–4 (2018). https://doi.org/10.1145/3239235.3267435

26. Martin, F., Johnson, M.: More efficient topic modelling through a noun only approach. In: Proceedings of the Australasian Language Technology Association Workshop 2015, pp. 111–115 (2015)

27. Mehrotra, R., Sanner, S., Buntine, W., Xie, L.: Improving LDA topic models for microblogs via tweet pooling and automatic labeling. In: Proceedings of the 36th International ACM SIGIR Conference on Research and Development in Information Retrieval, pp. 889–892 (2013). https://doi.org/10.1145/2484028.2484166

28. Miao, Y., Yu, L., Blunsom, P.: Neural variational inference for text processing. In: International Conference on Machine Learning, pp. 1727–1736. PMLR (2016)
29. Mikolov, T., Sutskever, I., Chen, K., Corrado, G.S., Dean, J.: Distributed representations of words and phrases and their compositionality. In: Advances in Neural Information Processing Systems, vol. 26 (2013)
30. Nalisnick, E., Smyth, P.: Stick-breaking variational autoencoders. In: Proceedings of the International Conference on Learning Representations (ICLR) (2017)
31. Nan, F., Ding, R., Nallapati, R., Xiang, B.: Topic modeling with Wasserstein autoencoders. arXiv preprint arXiv:1907.12374 (2019). https://doi.org/10.18653/v1/P19-1640
32. Newman, M., Girvan, M.: Finding and evaluating community structure in networks. Phys. Rev. E **69**(2), 026113 (2004). https://doi.org/10.1103/physreve.69.026113
33. Röder, M., Both, A., Hinneburg, A.: Exploring the space of topic coherence measures. In: Proceedings of the Eighth ACM International Conference on Web Search and Data Mining, pp. 399–408 (2015). https://doi.org/10.1145/2684822.2685324
34. Sakr, S., et al.: The future is big graphs: a community view on graph processing systems. Commun. ACM **64**(9), 62–71 (2021). https://doi.org/10.1145/3434642
35. Schofield, A., Mimno, D.: Comparing apples to apple: the effects of stemmers on topic models. Trans. Assoc. Comput. Linguist. **4**, 287–300 (2016). https://doi.org/10.1162/tacl_a_00099
36. Srivastava, A., Sutton, C.: Autoencoding variational inference for topic models. In: Proceedings of the International Conference on Learning Representations (ICLR) (2017)
37. Steyvers, M., Smyth, P., Rosen-Zvi, M., Griffiths, T.: Probabilistic author-topic models for information discovery. In: Proceedings of the Tenth ACM SIGKDD International Conference on Knowledge Discovery and Data Mining, pp. 306–315 (2004). https://doi.org/10.1145/1014052.1014087
38. Traag, V.A., Waltman, L., Van Eck, N.J.: From Louvain to Leiden: guaranteeing well-connected communities. Sci. Rep. **9**(1), 1–12 (2019). https://doi.org/10.1038/s41598-019-41695-z
39. Yang, K., Cai, Y., Chen, Z., Leung, H., Lau, R.: Exploring topic discriminating power of words in latent Dirichlet allocation. In: Proceedings of COLING 2016, the 26th International Conference on Computational Linguistics: Technical Papers, pp. 2238–2247 (2016)
40. Zhang, H., Chen, B., Guo, D., Zhou, M.: WHAI: Weibull hybrid autoencoding inference for deep topic modeling. In: 6th International Conference on Learning Representations (ICLR) (2018)
41. Zhao, H., Du, L., Buntine, W., Liu, G.: MetaLDA: a topic model that efficiently incorporates meta information. In: 2017 IEEE International Conference on Data Mining (ICDM), pp. 635–644 (2017). https://doi.org/10.1109/ICDM.2017.73

UMLS-Based Question-Answering Approach for Automatic Initial Frailty Assessment

Yashodhya V. Wijesinghe[1,2]([✉]), Yue Xu[1]([iD]), Yuefeng Li[1], and Qing Zhang[2]

[1] School of Computer Science, Queensland University of Technology, Brisbane 4000, Australia
`yashodhyavachila.wijesinghe@hdr.qut.edu.au`, `{yue.xu,y2.li}@qut.edu.au`
[2] The Australian e-Health Research Centre, CSIRO, Brisbane 4029, Australia
`qing.zhang@csiro.au`

Abstract. Frailty is the most problematic multidimensional geriatric syndrome among elderly population that leads to poor quality of life and increased risk of death. Adverse effects include an increased risk of hospitalisation and institutionalisation, poorer outcomes of post-hospitalisation, and higher mortality rates. A questionnaire-based frailty assessment is an effective way to achieve early diagnosis of frailty. However, most of the existing frailty assessment tools require face-to-face consultation. For elderly patients living in rural areas are more likely to struggle to access healthcare than a patient living in an urban or suburban area, and they have higher chance of catching diseases due to frequent hospital visits as most of them are vulnerable due to being immunocompromised. An automatic initial frailty assessment approach can minimise the impact of mentioned disadvantages and save clinical resources by avoiding unnecessary manual assessments. The objective of this paper is to propose an automatic initial frailty assessment approach which can quickly identify potential patients that require further frailty assessment by using patient's relevant clinical notes to answer Tillburg Frailty Indicator (TFI) questionnaire automatically. A phrase-based query expansion method is proposed to identify the most relevant phrases to the frailty assessment questionnaire based on UMLS ontology. The research shows the advantages of using UMLS based concepts as features in automatic initial frailty assessment using clinical notes. The research enables clinician to assess frailty automatically using medical data, reduces the frequency of face-to-face consultations and hospital visits, which is extremely beneficial during unusual or unexpected times such as COVID-19 pandemic.

Keywords: Frailty · UMLS · Question-answering

1 Introduction

Frailty is a physiological condition that can be found among the aging population which develops due to age-related deterioration of the physiological system.

L. A. F. Park et al. (Eds.): AusDM 2022, CCIS 1741, pp. 163–175, 2022.
https://doi.org/10.1007/978-981-19-8746-5_12

Frailty can be defined as "a state of vulnerability to poor resolution of homeostasis after a stressor event and is a consequence of the cumulative decline in many physiological systems during a lifetime" [1]. 5% of the population aged 60 and above suffers from frailty [2]. Early diagnosis of frailty can improve the primary care required by the elderly, as this would "minimizes the risk of developing the pre-frail state into frail state", control the negative impact on active lifestyle and reduce adverse effects such as recurrent hospitalization (contributes to increased health care cost), loss of independence and ultimately death [3–5]. The majority of the existing frailty assessment tools require face-to-face consultation to make the final assessment as they are based on set of questions and physical tests that require clinician opinion. The current COVID-19 global pandemic has a huge negative impact on elderly community due to being immuno-compromised. Therefore, regular hospital visitations and physical contact with health care workers can have negative impact on their health due increased exposure to diseases. Furthermore, some of the existing frailty indexes require assessor training and special and specialised equipment to measure frailty [6]. Therefore, the final decision can be subjective as the final decision depends on the level of training received by the assessor. The existing frailty index calculation approaches are costly and time consuming [6–8] due to human involvement. The proposed automatic initial frailty assessment approach can save limited clinical resources available by reducing the efforts spent on manual frailty assessment by automatically identifying elderly patients that require frailty assessment. The automated frailty assessment based on clinical notes can motivate toward proactive care for the elderly.

The increasing requirement to deliver high quality care with increased speed and at reduced cost have spiked an interest in automated Question-Answering systems (QA). In past, researchers developed QA systems to answer general questions related to medical domain using published medical literature and online resources such as MEDLINE articles, PubMed articles and so on. Examples of such systems are AskHERMES [9], MiPACQ [10], CLINIQA [11], MEANS [12] and so on. Most of the existing automated question answering systems in the medical domain are dependent on either knowledge graphs or medical documents. But none of them can be used for frailty assessment. Most of the existing general medical related QA systems apply either BM25 [9,12,13] or Lucene indexing tool to retrieve documents to answer general medical questions. BM25 is a ranking function which is used for document ranking based on query terms [13]. These existing systems used Unified Medical Language System (UMLS) to annotate the questions. Moreover, most of the existing approaches focus only on terms when retrieving documents and BM25 tends to fail as the meaning of words is not considered, thereby misrepresenting the documents that carry the same information but use different words [14]. In this research, we propose a word embedding based approach to identify answers to specific questions of the assessment based on word meaning and word matching as well. Word embedding is techniques where the meaning of individual words is represented as real-valued vectors in a predefined vector space [15].

The Electronic Health Record (EHR) is digitized version of systematically collected medical information. EHRs provide a comprehensive medical history which can be used in automated initial frailty assessment. This research focuses on developing text mining methods to automatically identify facts or information in clinical notes available in EHRs to answer questions in Tilburg Frailty Indicator (TFI) questionnaire to assess patients' frailty. These clinical notes are written by clinicians. As the input to the proposed approach the set of clinical notes belonging to a patient is passed which in turn generates the TFI frailty score of that particular patient as the output. The proposed approach enables to initially screen the elderly patients for frailty without the restrictions of physical, geographical or environmental conditions and also reduce time or financial cost caused by face-to-face consultations.

In this paper, Sect. 2 discusses the state-of-art approaches related to the proposed research model. The proposed approach is described in Sect. 3. The experimental results are elaborated under Sect. 4. The Sect. 5 analyses results. The Sect. 6 concludes the paper.

2   Related Work

With the heightened interest in applying data science to medical domain, many QA systems have been developed by researchers such as AskHERMES [9], MiPACQ [10], MEANS [12] and so on. AskHERMES is a question-and-answer system for doctors, which can answer complicated clinical practice questions. These medical domain related QA systems answer general medical questions using MEDLINE abstracts, e-medicine documents, clinical guidelines, full text research articles and Wikipedia documents. These approaches use UMLS to annotate questions as well as documents with UMLS entity types and to expand query terms. Then BM25 [16] is used for document retrieval. Saleh et al. proposed a method for term selection for medical information retrieval where query expansion is applied only when the model predicts a score for a candidate term using Kullback-Leiber Divergence (KLD) which allows to expand queries with strongly related terms only [17]. The researches have used multiple documents sources such as Wikipedia articles and PubMed abstracts for query expansion.

UMLS is an open-source biomedical ontology, developed by US National Library of Medicine [18]. The UMLS compromises of approximately 3.2 million biomedical concepts [19]. A concept represents a set or class of entities or 'things' within a domain. In today's world, research and healthcare practices generates vast amount of digital content. Managing and analysing this data can be challenging due to size, ambiguity, complexity and heterogeneity of this data [20]. UMLS can be used to deal with the challenging.

Tilburg Frailty Indicator (TFI) is a self-reported user-friendly questionnaire developed by Gobbens et al. [21, 22] to measure frailty in the elderly. The questionnaire is used by general practitioners, nurses, social workers, and other health care workers. Therefore, questionnaire requires face-to-face consultation between clinician and the patient is required to diagnose frailty. The questionnaire consists of 15 questions. The initial query phrases are identified manually based on the keywords of the question. Table 1 depicts the questions, identified initial query phrases and frailty score for each question [21].

Table 1. TFI questions and frailty score for each question

Question number	Question	Initial query phrase	Frailty score for the answer
1	Do you feel physically healthy?	Physically healthy	Yes: 0 No: 1
2	Have you lost a lot of weight recently without wishing to do so?	Lost weight	Yes: 1 No: 0
3	Do you experience problems in your daily life due to: difficulty in walking?	Difficulty walking	Yes: 1 No: 0
4	Do you experience problems in your daily life due to: difficulty maintaining your balance?	Difficulty balancing	Yes: 1 No: 0
5	Do you experience problems in your daily life due to: poor hearing?	Poor hearing	Yes: 1 No: 0
6	Do you experience problems in your daily life due to: poor vision?	Poor vision	Yes: 1 No: 0
7	Do you experience problems in your daily life due to: lack of strength in your hands?	Lack hands strength	Yes: 1 No: 0
8	Do you experience problems in your daily life due to: physical tiredness?	Physical tiredness	Yes - 1 No - 0
9	Do you have problems with your memory?	Memory problem	Yes: 1 No: 0
10	Have you felt down during the last month?	Felt down	Yes: 1 No: 0
11	Have you felt nervous or anxious during the last month?	Anxious or nervous	Yes: 1 No: 0
12	Are you able to cope with problems well?	Cope with problems	Yes: 0 No: 1
13	Do you live alone?	Live alone	Yes: 1 No: 0
14	Do you sometimes miss having people around you?	Miss people	Yes: 1 No: 0
15	Do you receive enough support from other people?	Support from people	Yes: 0 No: 1

With respect to the original TFI questionnaire, 11 of the questions in the questionnaire have a dichotomous answer format (yes/no) and 4 questions have a threefold answer format (yes/no/sometimes) for questions, 9, 10, 11, and 14. For question 9 the answer 'sometimes' is recoded into 'no' and for questions 10,

11, 14 'sometimes' is recoded into 'yes' [21]. Therefore, questions in the TFI requires binary answer. In this paper for the purpose of simplicity, we do not consider the option 'sometimes' when answering the question. For the questions 1, 12 and 15 the answer 'yes' is scored with 0 and 'no' with 1. For the remaining questions the answer 'yes' is scored as 1 and 'no' with 0. The question scores are summed into a frailty score ranging from 0–15. Patients with a total TFI score ≥ 5 are considered to be frail. The TFI allows elderly to be diagnosed as frail or not-frail based on the threshold value.

3   The Proposed Approach

3.1   Discovery of UMLS Based Concepts

The research focuses on developing text mining methods to determine frailty of elderly patients by answering TFI questionnaire using UMLS. The main goal of the research is to develop text mining methods to identify relevant UMLS concepts that could be used for answering each question in TFI questionnaire. To better represent the main point of each question an initial phrase from each question is manually identified based on the keywords of the question.

A phrase can be defined as a small group of words that are standing together as a conceptual unit where the order of the words is important. As an example, 'poor vision' is a phrase to describe a patient's vision condition. These initial phrases are used to find relevant UMLS concepts for answering the questions in the TFI questionnaire instead of using individual terms. The terma are ambiguous and less discriminative compared to phrases. There are few problems in identifying relevant clinical notes for answering the questionnaire using only the initial query phrase. One problem is that physicians may not use the same phrases as the initial phrases in the questionnaire to describe the patients' conditions in various notes. As an example, 'poor vision' is an initial query phrase identified from a question, whereas physician may use a phrase 'loss of sight' or 'blind' to describe the comorbidity. In this case, some relevant clinical notes which contain different phrases from the initial phrases in the questionnaire would not be identified for answering the questionnaire questions even though they actually contain the answer to the questions. Therefore, generating a set of relevant UMLS concepts to expand the initial query phrases has become a key issue.

The proposed system consists of two main components, query expansion and frailty assessment. With the query expansion, we can add relevant phrases into the query so that it is possible to retrieve the clinical notes that can be used to answer the question even the clinical notes do not contain the initial phrases. The Fig. 1 illustrates a high-level architecture diagram for the proposed system.

As mentioned above, there are well-established domain ontologies in medical and healthcare domain such as UMLS. Using ontology allows us to identify domain specific phrases. The phrases identified in UMLS may be different from the initial phrases in the questionnaire but share the same concepts. For example, the initial query phrase 'poor vision' is represented in UMLS as 'blurred vision',

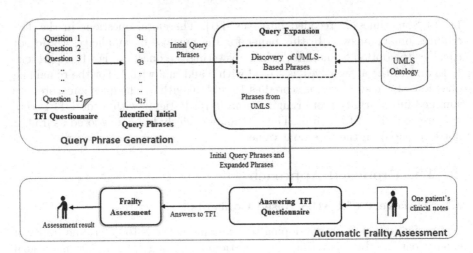

Fig. 1. The overview architecture diagram of UMLS-based automatic initial frailty assessment

'low vision', 'myopia' and so on. In order to capture relevant phrases to the initial query phrases in the questionnaire, this paper proposes to expand the queries in the questionnaire based on medical ontology UMLS.

Individual terms may lead to ambiguity as a term such 'poor' can be used to describe any medical scenario while the term 'vision' doesn't give an accurate description of patient's ability to see. In order to find meaningful phrases, we need to capture the meaning of individual terms based on which to form meaningful phrases. To achieve this, we propose to assess the overlapping and also the meaning similarity between the initial query and the concepts. In this paper, word embeddings are generated to represent individual words as word embedding captures the semantic meaning of each word. These word embeddings will be used to generate embeddings to represent phrases.

3.2 UMLS-Based Concept Selection Algorithm

From the 15 questions in the TFI questionnaire we can extract 15 initial queries each of which corresponds to one question. Let Q be a set of 15 initial queries denoted as $Q = \{q_1, q_2, \ldots, q_{15}\}$, q_j is a query phrase containing a set of words, $q_j = \{w_1, w_2, \ldots, w_{|q_j|}\}$. Let C be a set of all the medical concepts in UMLS. The Rest Application Programming Interface (API) 'search' of UMLS is used to return all the relevant concepts to q from UMLS. Let $concept\_UMLS(q)$ denote the set of relevant concepts to q, $concept\_UMLS(q) = search(q)$, $concept\_UMLS(q) \subset C$. We can further determine the relevant concepts to the query phrase q by examining the similarity between q and the related concepts. The similarity between q and the concepts in $concept\_UMLS(q)$ can be calculated based on their word embeddings. The word embedding for a phrase or a concept q is defined below, where $vec(w)$ is the word embedding of word w.

$$vec(q) = \frac{\sum_{w \in q} vec(w)}{|q|} \qquad (1)$$

For an initial query q, its UMLS related phrases are defined as $phrase\_UMLS(q) = \{c \mid c \in concept\_UMLS(q), c \cap q \neq \{\}, len(c) \leq x, sim(vec(q), vec(c)) > y)\}$. The x is the maximum length of a UMLS based phrase. According to the existing literature the extremely lengthy phrases introduce noise [23]. Therefore, length of the selected phrases is restricted and x denotes the maximum length of a UMLS based query phrase. The y is the similarity threshold value. The UMLS-based phrases in $phrase\_UMLS(q)$ are used to expand the query q. The list of expanded query phrases along with the initial query phrase is defined as $IU(q) = \{q\} \cup phrase\_UMLS(q)$.

Algorithm 1: Discovery of UMLS Based Phrases

Input: Q : A list of query phrases identified from questions
C : A set of concepts of UMLS
x, y : threshold values for length and similarity respectively
Output: $phrase\_UMLS$:UMLS-based phrases

1 for $q \in Q$
2   vec $vec(q) = \frac{\sum_{w \in q} vec(w)}{|q|}$
3      $concept\_UMLS(q) := search(q)$ //'search is a function available in UMLS Rest API'
4      $phrase\_UMLS(q) := \{\}$ // List of selected UMLS based query phrases or query q
5   for each $c \in concept\_UMLS(q)$
6   if $len(c) \leq x$
7              $vec(c) := \frac{\sum_{w \in c} vec(w)}{|c|}$
8   if $sim(vec(q), vec(c)) \geq y$ // y
9                  $phrase\_UMLS(q) = phrase\_UMLS(q) \cup \{c\}$
10 return $phrase\_UMLS$

As the input to the Algorithm 1, Q represents the list of initial query phrases identified from each question. Line 2 in the algorithm demonstrate representation of query phrase using word embedding. In Line 3, using API function 'search' a set of related UMLS concepts are returned from the ontology. From line 5 to 9 elaborates discovery of UMLS concepts related to input query phrase. The Once the ontology returns list of related concepts, in the line 6 length of the concept is checked, then in the line 7 selected concept is represented using word embedding. In the line 8 of the algorithm, similarity between UMLS concept and initial query phrase is calculated using cosine similarity and the concept that satisfies the minimum similarity threshold value is selected as a UMLS based concept. And the selected UMLS based concepts are used to expand the initial query.

3.3   Answering TFI Questionnaire Using UMLS-Based Concepts

When answering a question in the TFI questionnaire, if any of the query phrases related to the question is available in a clinical note, the similarity between all the query phrases and the initial query phrase which are available in the clinical note will be assessed to decide whether this particular clinical note can be used to answer this question or not. A clinical note can contain multiple query phrases with varying degree of similarity to the initial query phrase. A simple way is to calculate the average similarity over all the available phrases to the initial query phrase. If the average similarity is equal or greater than 0.5, which means the query phrases that are in the clinical note is similar to the manually selected initial query phrase, then this clinical note is considered as a positive scenario. Therefore, the answer to the particular question is considered 'Yes' as this allows us to retrieve notes that can be used to answer the question. However, if the similarity between the initial query phrase and the query phrase available in the note is less than 0.5, it will be considered as a negative case. Therefore, the question will be answered 'No'. Additionally, if any notes do not contain any of the query phrases, then the question will be answered as 'No'. The hypothesis behind this is that if patient does not suffer from a particular ailment, then the clinician would not be writing anything regarding that particular ailment.

3.4   Frailty Assessment

The answers given to the questions in the questionnaire are scored and summed into a frailty score ranging from 0–15 as there are 15 questions in the questionnaire. In the questionnaire, if the questions 2 to 11, 13 and 14 are answered as 'yes', then score 1 is assigned, whereas the answer 'no' for these questions would be scored as 0 [21]. For the questions 1,12 and 15, if the answer is 'yes', then they will be scored as 0, whereas if the answer is 'no' then the assigned value would be 1. Once all the questions are scored as either 1 or 0, the sum of the question scores will be used to determine the initial frailty. If the sum is greater than or equal 5, then the patient is considered frail, otherwise not frail.

4   Experiment and Results

4.1   Dataset

As the dataset of this research, we are using a set of hospital records belonging to a cohort of 693 elderly patients. The data consist of demographics, medication, laboratory test results and clinical notes. In this research, we will be using only the clinical notes which are written by clinicians. The clinical notes consist of multiple types of notes. In this study we use only 'admission notes', 'discharge notes', 'social history', 'mobility', 'activities of daily life (ADL)' and 'discharge care plan' to answer TFI questionnaire automatically. Each record in the dataset is a patient's one admission data. When answering TFI questionnaire for one patient, patient's clinical notes related to a particular admission would be considered as the input to search for answers. The feature 'md_frailty' in the dataset is used as the ground truth for evaluating the accuracy of the automatic frailty assessment. The value of the feature 'md_frailty' is CFS score. A patient is considered as frail if the CFS score is larger than or equal to 4 [24].

All the text documents in the collection were converted to lowercase and each word in the document is considered as a token. The stop words, roman numerals, numbers, days of the week, months, seasons and so on were removed from the tokenized documents. Once all the unnecessary terms are removed, the remaining tokens are stemmed to have a common root.

4.2 Experiment Settings

The research proposes 2 query expansion approaches to retrieve clinical notes to answer TFI questionnaire for a patient, which are defined below:

- IQ: This model uses the manually identified initial query phrase to retrieve clinical notes to answer TFI questionnaire.
- IQ_UMLS: This model uses the initial query phrase and the query phrases extracted from the UMLS. The approach proposed in Sect. 3.2 is used to retrieve related phrases from the UMLS. This approach can be used to evaluate the effectiveness of the query phrases derived from the ontology when answering frailty questionnaire.

In the experiments, the maximum length of the query phrase generated by UMLS ontology x is set to 3 [23], while the minimum similarity between UMLS based phrase and initial query phrase y is set to 0.5.

In order to evaluate the performance of the proposed methods, we compare our methods against existing methods;

- Term-based Model [17]: In this approach, the terms in the query phrases are used to retrieve clinical document. As an example, the terms 'poor' and 'hearing' in the query phrase 'poor hearing' are used to retrieve clinical notes to the related question.
- AskHerms [10]: This model was proposed using BM25 to score the terms, based on which to retrieve documents to answer clinical questions. The relevant documents are retrieved from MEDLINE abstracts, PubMed Central full-text articles, eMedicine documents, clinical guidelines and Wikipedia articles when answering such as 'What is the treatment (or therapy) of polymenorrhea for a 14-year-old girl?'. The documents were retrieved based on the question-document similarity which is measured based using BM25 similarity.

4.3 Results

We compared the performance of frailty assessment using the three proposed query expansion methods. Each patient in the dataset is used as a test case. The frailty assessment result to a patient produced by our methods is compared against the ground truth in the dataset. Metrics Precision, Recall, F1 Score and Accuracy are used to measure the performance.

Table 2, depicts the performance results for the two proposed methods. In Table 2 it can be observed that the approach IQ_UMLS gives the best performance. As described in Sect. 3, methods IQ may miss relevant clinical notes as these approaches may not generate some possible phrases that were used by clinicians to describe a clinical condition. Therefore, method IQ_UMLS which uses the initial query phrase along with UMLS based concepts leads to better performance.

Table 2. Frailty assessment results of the proposed methods

Proposed clinical notes selection approach	Precision %	Recall %	F1 Score %	Accuracy %
IQ	63.43	**97.58**	76.91	64.17
IQ_UMLS	73.43	96.77	83.50	73.18

The Table 3 below depicts the comparison results between IQ_UMLS, the best approach among the proposed approaches, against the baseline query expansion approaches [9,17]. It can be observed in Table 3 that IQ_UMLS outperforms the baseline models when assessing the frailty.

Table 3. Comparison between proposed model and baselines for frailty assessment

The best proposed model and baselines	Precision %	Recall %	F1 Score %	Accuracy %
Terms-Based [17]	40.21	90.20	55.62	40.71
AskHerms [9]	47.20	91.53	62.28	47.17
IQ_UMLS	**73.43**	**96.77**	**83.50**	**73.18**

If no relevant clinical notes for a question can be found, then it is assumed that the question is answered negatively. Compared to the baselines, the proposed model IQ_UMLS can retrieve more relevant clinical notes as UMLS is medical ontology. Hence, the identified UMLS based concepts are domain specific. This in turn leads to increase performance compared to baselines.

5 Discussion

Currently, the developed phrase-based question-answering for automatic initial frailty assessment is in the early stage. TFI and Edmonton Frailty scale are commonly used frailty questionnaires. These two fraily assessments consist of set of pre-defined questions unlike other frailty assessment tools (CFS, Rockwood Mitnitski Frailty Index etc.). The pre-defined questions make automatic frailty screening using clinical notes become possible.

When answering the TFI questionnaire, we have considered all types of clinical notes belonging to a patient as this gives a better overall understanding about the patient's condition. Considering only a few types of notes such as admission notes, physician notes and so on could lead to loss of crucial information required to answer the questionnaire.

The pilot evaluation results showed that the proposed approach outperforms the baseline models for query expansion which in turn leads to increased performance of initial frailty assessment. One of the major factors that contributes toward higher F1 score and accuracy is the identified note-based query phrases which can represent the question that's being answered. The identified note-based query phrases are obtained from patients' clinical notes, and hence are more representative of patients' clinical conditions. Furthermore, using phrases instead of using terms increases the chance to find relevant notes to answer questions because phrases are more semantically similar to the identified initial query phrase of each question.

6   Conclusion

Mainly, this research focuses on developing an automatic frailty screening approach to answer the questions in a frailty questionnaire based on patients' clinical notes and medical ontology UMLS to identify elderly patients who require further face-to-face frailty assessment. An automatic initial frailty assessment can reduce face to face interaction between patients and clinicians when assessing frailty as this can reduce can reduce the number of hospital visitations, save hospital resources, require less time and can make life more easier for elderly, especially beneficial to remote areas or during special/unusual periods such as the COVID-19 pandemic. To answer the frailty questionnaire, an initial phrase was identified to represent each question. Then the initial query phrase was expanded by using the proposed query expansion approach to identify meaningful, related and semantically similar query phrases based on UMLS medical ontology. and clinical notes. The expanded query phrases are then used to retrieve the most relevant notes to answer the questions in the questionnaire. The experimental results show that the proposed approach outperforms the baseline approaches, which indicates that, the expanded query phrases by the proposed model are more representative and meaningful than the baseline approaches. The main limitation of this approach is that some of the specific or uncommon or indirect phrases used by clinicians may be not captured and sometimes the answer to a specific question can be inferred based on the answers given to other questions, e.g., in the question 15, the clinical notes contain phrases such as 'son pays bill' which indicates that the patient receives support from others, but these kind of phrases were not captured by our query expansion approach. Because of this, some of the positive scenarios related to question 15 were identified as negative scenario. Additionally, the proposed approach need to be tested on multiple clinical datasets as in order to confirm that same level of accuracy can be achieved. This limitation will be addressed in our future work to deal with the indirect relationship between expanded phrases of a particular question and other questions. However, all the questions in the TFI can be answered using clinical notes. In conclusion, the study enables the clinician to assess frailty of elderly patients based on the patients' medical data without requiring physical consultations or hospital visitations.

Acknowledgements. The authors wish to thank Cheng Hwee Soh, Dr. Andrea Britta and Dr. Lim Kwang from University of Melbourne for providing the dataset for this research.

References

1. Clegg, A., Young, J., Iliffe, S., Rikkert, M.O., Rockwood, K.: Frailty in elderly people. Lancet **381**(9868), 752–762 (2013)
2. Malmstrom, T.K., Miller, D.K., Morley, J.E.: A comparison of four frailty models. J. Am. Geriatr. Soc. **62**(4), 721–726 (2014)
3. de Gelder, J., et al.: Predicting adverse health outcomes in older emergency department patients: the APOP study. Neth. J. Med. **74**(8), 342–352 (2016)
4. Rockwood, K., et al.: A global clinical measure of fitness and frailty in elderly people. Can. Med. Assoc. J. **173**(5), 489–495 (2005)
5. Sternberg, S.A., Schwartz, A.W., Karunananthan, S., Bergman, H., Clarfield, M.: The identification of frailty: a systematic literature review. J. Am. Geriatr. Soc. **59**(11), 2129–2138 (2011)
6. Dent, E., Kowal, P., Hoogendijk, E.O.: Frailty measurement in research and clinical practice: a review. Eur. J. Int. Med. **31**, 3–10 (2016)
7. Newgard, C.D., Zive, D., Jui, J., Weathers, C., Daya, M.: Electronic versus manual data processing: evaluating the use of electronic health records in out-of-hospital clinical research. Acad. Emerg. Med. Off. J. Soc. Acad. Emerg. Med **19**(2), 217–227 (2012)
8. Pavlović, I., Kern, T., Miklavcic, D.: Comparison of paper-based and electronic data collection process in clinical trials: costs simulation study. Contemp. Clin. Trials **30**(4), 300–316 (2009)
9. Cao, Y., et al.: AskHERMES: an online question answering system for complex clinical questions. J. Biomed. Inform. **44**(2), 277–288 (2011)
10. Cairns, B.L., et al.: The MiPACQ clinical question answering system. In: AMIA Annual Symposium Proceedings, pp. 171–180 (2011)
11. Zahid, M., Mittal, A., Joshi, R.C., Atluri, G.: CliniQA: a machine intelligence based clinical question answering system. arXiv preprint arXiv:1805.05927 (2018)
12. Abacha, A.B., Zweigenbaum, P.: Means: a medical question-answering system combining NLP techniques and semantic web technologies. Inf. Process. Manag. **51**(5), 570–594 (2015)
13. Amati, G.: BM25. In: Liu, L., Özsu, M.T. (eds.) Encyclopedia of Database Systems, pp. 257–260. Springer, Boston (2009). https://doi.org/10.1007/978-0-387-39940-9_921
14. Gurulingappa, H., Toldo, L., Schepers, C., Bauer, A., Megaro, G.: Semi-supervised information retrieval system for clinical decision support. In: TREC (2016)

15. Goldberg, Y.: Neural network methods for natural language processing. Synth. Lect. Hum. Lang. Technol. **10**(1), 1–309 (2017)
16. Robertson, S., Zaragoza, H.: The probabilistic relevance framework: BM25 and beyond. Found. Trends Inf. Retr. **3**, 333–389 (2009)
17. Saleh, S., Pecina, P.: Term selection for query expansion in medical cross-lingual information retrieval. In: Azzopardi, L., Stein, B., Fuhr, N., Mayr, P., Hauff, C., Hiemstra, D. (eds.) ECIR 2019. LNCS, vol. 11437, pp. 507–522. Springer, Cham (2019). https://doi.org/10.1007/978-3-030-15712-8_33
18. Bodenreider, O.: The unified medical language system (UMLs): integrating biomedical terminology. Nucleic Acids Res. **32**(Database issue), D267–D270 (2004)
19. Mohan, S., Li, D.: MedMentions: a large biomedical corpus annotated with UMLs concepts. arXiv preprint arXiv:1902.09476 (2019)
20. Ferreira, J.D., Teixeira, D.C., Pesquita, C.: Biomedical Ontologies: Coverage, Access and Use, pp. 382–395. Academic Press, Oxford (2021)
21. Gobbens, R.J., van Assen, M.A., Luijkx, K.G., Wijnen-Sponselee, M.T., Schols, J.: The Tilburg frailty indicator: psychometric properties. J. Am. Med. Dir. Assoc. **11**(5), 344–355 (2010)
22. Gobbens, R.J.J., van Assen, M.A.L.M.: The prediction of quality of life by physical, psychological and social components of frailty in community-dwelling older people. Qual. Life Res. **23**(8), 2289–2300 (2014). https://doi.org/10.1007/s11136-014-0672-1
23. Koehn, P., Och, F.J., Marcu, D.: Statistical phrase-based translation. In: Proceedings of the 2003 Conference of the North American Chapter of the Association for Computational Linguistics on Human Language Technology, vol. 1, pp. 48–54. Association for Computational Linguistics (2003)
24. Rockwood, K., Theou, O.: Using the clinical frailty scale in allocating scarce health care resources. Can. Geriatr. J. **23**(3), 210–215 (2020)

Natural Language Query for Technical Knowledge Graph Navigation

Ziyu Zhao[1,2]([✉]), Michael Stewart[1,2], Wei Liu[1,2], Tim French[1,2],
and Melinda Hodkiewicz[1,3]

[1] UWA NLP-TLP Group, The University of Western Australia, Perth, Australia
`ziyu.zhao@research.uwa.edu.au`,
`{michael.stewart,wei.liu,tim.french,melinda.hodkiewicz}@uwa.edu.au`
[2] School of Physics, Mathematics and Computing, The University of Western
Australia, Perth, Australia
[3] School of Engineering, The University of Western Australia, Perth, Australia
`https://nlp-tlp.org/`

Abstract. Technical knowledge graphs are difficult to navigate. To support users with no coding experience, one can use traditional structured HTML form controls, such as drop-down lists and check-boxes, to construct queries. However, this requires multiple clicks and selections. Natural language queries, on the other hand, are more convenient and less restrictive for knowledge graphs navigation. In this paper, we propose a system that enables natural language queries against technical knowledge graphs. Given an input utterance (i.e., a query in human language), we first perform Named Entity Recognition (NER) to identify domain specific entity mentions as node names, entity types as node labels, and question words (e.g., `what`, `how many` and `list`) as keywords of a structured query language before the rule-based formal query constructions. Three rules are exploited to generate a valid structured formal query. The web-based interactive application is developed to help maintainers access industrial maintenance knowledge graph which is constructed from text data.

Keywords: Natural language query · Rule-based semantic parsing · Technical knowledge graph

1 Introduction

Using structured query languages (e.g., GraphQL[1], Cypher[2] to navigate through large Knowledge Graphs (KGs) such as those extracted from unstructured texts, is a challenging task for end users who have no coding experience. *Natural language queries* aim to allow users to interact with such data in a human language (such as English), rather than relying on form controls such as drop-down lists and check-boxes, which have limited expressiveness.

---

[1] `GraphQL`: https://graphql.org/learn/.

[2] `Neo4j Cypher`: https://neo4j.com/docs/cypher-manual/current/).

© The Author(s), under exclusive license to Springer Nature Singapore Pte Ltd. 2022
L. A. F. Park et al. (Eds.): AusDM 2022, CCIS 1741, pp. 176–191, 2022.
https://doi.org/10.1007/978-981-19-8746-5_13

The advancement in natural language processing makes natural language queries very appealing for many applications, such as knowledge base queries [42] and text to SQL over relational database [19]. It provides a convenient way to interact with databases without requiring one to be conversant in structured query languages and familiar with the databases' schema for constructing a valid query. In addition, natural language queries, as an essential step to voice-enabled digital assistants, pave the way to hands-free navigation of KGs in industry.

The fundamental challenge to supporting a natural language queries is how to translate a natural language utterance to a machine readable and interpretable formal query language, which is known as *semantic parsing* [32]. Given a question in human language, a semantic parsing model predicts (based on pre-trained neural network models) or generates (based on rules) a machine-readable structured query that is then executed against a relational database or a graph database. Research has shown that natural language queries can be parsed into Prolog [29], lambda(λ) calculus [41], SQL [19] for relational databases, SPARQL [37] and Cypher [12] for graph databases. Table 1 shows examples of natural language utterance and their respective structured formal queries.

Table 1. The examples of natural language utterance and their structured semantic representation

Input:	What is the largest city in Texas?
Prolog Query [29]:	`answer(C,largest(C,(city(C),loc(C,S),const(S,stateid(texas)))))`
Input:	What is the largest state?
λ-calculus Query [41]:	`argmax(λx.state(x), λx.size(x))`
Input:	Return me the number of papers on PVLDB
Table:	Journal||Publication||Author||Conference||Domain||Organization||Cite||
SQL Query [19]:	`SELECT COUNT(DISTINCT t2.title) FROM Publication AS T2 JOIN Journal AS T1` `ON T2.JID = T1.JID WHERE T1.name = "PVLDB"`
Input:	Who voiced meg on family guy?

SPARQL Query [37]:	`PREFIX ns: <http://rdf.freebase.com/ns/>`
	`SELECT ?x`
	`WHERE { ns:m.035szd ns:tv.tv_character.appeared_in_tv_program ?y0 .`
	`?y0 ns:tv.regular_tv_appearance.actor ?x ; ns:tv.regular_tv_appearance.series ns:m.019nnl ;`
	`ns:tv.regular_tv_appearance.special_performance_type ns:m.02nsjvf .}`

Input:	Shows me potential fraud detection rings

Cypher Query [12]:	`MATCH (accHolder:AccountHolder)-[:HAS]->(pInfo)`
	`WHERE pInfo:SSN OR pInfo:PhoneNumber OR pInfo:Address`
	`WITH pInfo, collect(accHolder.uniqueId) AS accountHolders, count(*) AS fraudRingCount`
	`WHERE fraudRing > 1`
	`RETURN accountHolders, labels(pInfo) AS personalInformation, fraudRingCount`

Recently, semantic parsing models [10,16] based on neural networks components (e.g., LSTM [15], CNN [18], RNN [23] and GNN [24]) and pre-trained embeddings (e.g., GloVe [20], ElMo [21] and BERT [9]) have achieved impressive results in general non-domain specific areas such as Freebase[3] and Wikidata[4]

---

[3] **Freebase:** https://developers.google.com/freebase.
[4] **Wikidata:** https://www.wikidata.org.

One can efficiently train neural networks with gradient-based optimization algorithms, given sufficient natural and structured query pairs. Unfortunately, some technical issues still prevent neural based sequence-to-sequence learning from being suitable for industrial applications. First, the language used is terse without correct grammatical structure, and Part of Speech tagging tools thus fail to perform. Second, the vocabulary used now has domain specific meaning, which makes general purpose word embeddings ineffective. The third and the most limiting factor is the lack of quality parallel training data and the difficulty in interpreting these neural models, making it even more challenging for model transfer. To support practical industry applications, such as inspection and diagnostics, predictive and prognostic maintenance, in this work, we propose a system that combines neural networks and pre-defined domain rules for semantic parsing, to support natural language queries in English in low-resource technical domains with limited training data. In particular, we demonstrate an application of the proposed pipeline in the maintenance domain that requires explainable model, and the contributions are as follows.

1. It provides reasonably good natural language access to technical knowledge graph with precise answers to questions;
2. Handles basic morphological use of entity-level normalization (aka node normalisation) and correct spelling errors.
3. Handles three variants of natural language queries: list type queries, aggregation queries and complex queries, of which aggregation and complex queries are particularly difficult to express using traditional HTML form controls.

2   Related Work

Traditional natural language queries over structured data are rule-based. Typically, syntax analysis is performed firstly to obtain the chunks or syntactic structures for the given input utterance, and next a set of manual lexicons and translation rules are applied. Later, empirical or corpus-based systems replace hand-curated rules with learned rules obtained automatically by training models over parallel corpus. Along with the success and wider adoption of machine learning, neural-based natural language interfaces have become a trend. In this section, we introduce the related works in classical rule-based methods and modern neural-based methods.

Rule-Based Natural Language Interface. One of the earliest *rule-based* works that translate utterance in natural language into a formal query was proposed by Kirsch [17]. It was designed to verify if natural language statements were correct for a given image. A natural language statement was parsed to a syntactic tree using a constituent grammar, and the rules of formation translate the syntactic tree to first order predicates. The learned formal logic's expression was executed with the objective to obtain either true, vacuously true (true with a condition) or false result. Another example is LUNAR [33]. The system supports users in asking questions about lunar geology. It carries out syntax analysis to have input's grammatical structure, and then transforms syntactic fragments using rules to structured semantic representation. Other rule-based systems [14,31] either adopted pattern matching strategies or sentences' syntactic tree as an intermediate representation to map to corresponding structured query based on predefined rules. Since syntactic analysis is a small part of the overall problem of natural language understanding and the reliance on predefined rules or lexicons might limit models from scaling and transferring to different domains. Some researchers proposed to train models on annotated datasets of natural language queries paired with logical-forms. CHILL [40] is one of the first systems that used corpus-based methods. The *corpus-based* methods typically learn lexicalized mapping rules to construct a candidate set of semantic representations. It provides a natural language interface to the U.S. geography database. In Table 1, the first example shows a query from the created geography dataset. Rather than containing a parser in the contents of the stack and the remaining input buffer, CHILL used its training examples to learn search-control rules within a logic program representing a overly-general shift-reduce parser written in Prolog.

Neural Natural Language Interface. Another popular line of research regarding this topic is based on *neural networks* in a sequence-to-sequence manner. A variety of this kind of models have been proposed to learn natural language sentences paired with their semantic meaning representation [10,25]. In this formalism, a significant large training corpus is required in order to train a neural network with an objective function and optimisation algorithms to learn the model's parameters. During testing, the trained model predicts the results of input examples. Text-to-SQL research has recently become popular due to the success of large end-to-end neural models for conditional text generation [3] and the availability of large datasets, such as WikiSQL [43], Spider [39] and CoSQL [38]. Other popular text to formal language datasets are GEO [40] which is about US geography of natural language questions paired with corresponding Prolog database queries, and ATIS [8] which converts natural language utterance to λ-calculus[5] (a simple notation for functions and application) queries.

Natural Language Query Against Graph Databases. Graph databases are useful for representing, organising and analysing data. For example, users may aggregate over complex interconnected information in knowledge graphs (KGs). Other applications include such as question answering [7], recommendation systems [30] and fraud detection [22]. Semantic information from KGs can be used to enhance search results in natural language interfaces against graph databases. One of the famous question answering systems is Watson [11] developed by IBM. It is built using knowledge bases, such as YAGO[6] [27], and DBpedia[7] [2], showing the value of KGs. Other semantic reorientation queries are based on KG queries [36]. Many researches use Freebase [6] as a source

---

[5] λ-calculus: https://plato.stanford.edu/entries/lambda-calculus/.

[6] YAGO: https://yago-knowledge.org.

[7] DBpedia: https://www.dbpedia.org/resources/knowledge-graphs/.

of knowledge for semantic parsing [4] and question answering [34]. The fourth example in Table 1 shows a query through a Freebase SPARQL endpoint. The last example uses Cypher query to retrieve the potential fraud detection rings.

Named Entity Recognition. Knowledge graphs offer a widely used format for representing informative information in computer-processable form. Named entity recognition (NER) is a central part to mining KGs from natural language text. Traditional NER systems include two sub-tasks: 1) identify entity boundaries, and 2) classify detected named entities into a set of labels (typically less than 10 of coarse types). The first shared task on NER was proposed by Grishman et al. [13] when the term `Named Entity` was first used. It aims to assign semantic types (e.g., `person`, `location`, `organization`) to the mentioned entities in the presence of their context. Neural NER has become the main trend thanks to the recent advances in neural architectures [35]. Unlike everyday English in natural language, technical language, such as maintenance work orders in the field of industrial maintenance and medical descriptions in public health domain, presents unique challenges, including: the varying styles of different maintainers and organisations; domain jargon; inconsistent and ambiguous use of words; mis-spelling; short hand notation; and the lack of syntactic grammatical structure. Additionally, due to confidentiality concerns, labelled training data for technical text processing is rarely available.

The work that is most closely related to ours is proposed by Sun et al. [28], which do not have node normalize process. Instead, they trained NER modules directly for graph components, namely graph nodes and graph edges. The work presented in this paper is the first Text-to-Cypher for maintenance work orders (MWO), and there is currently no publicly available parallel annotated dataset, nor a benchmark for maintenance knowledge graph. Therefore, we use the evaluation metrics of NER to measure the performance of semantic parsing for Text-to-Cypher, and show that the accuracy of the semantic parsing over graph database in terms of accuracy is improved from the level of performance of ensemble NER modules used in our approach. End-to-end neural network based models are playing a big role in the overall natural language understanding, especially in automatic feature engineering for text data. However, in the case of no data or limited data, our proposed rules are more practical, interpretable and controllable, and so is the proposed ensemble NERs for the MWO semantics understanding, by taking into account entity types' information.

3   Approach

Figure 1 illustrates the overall architecture for supporting queries of large technical knowledge graphs in a natural language. An interactive web interface and is used to collect users' natural language queries and respond with sub-graph and a tabular visualisation and tabular display of query results. The pipeline in the back-end translates utterances to structured formal query. It is composed of three main modules, namely, an neural NER Ensemble module, node normalisation and query writer (or rule-based structured formal query generator).

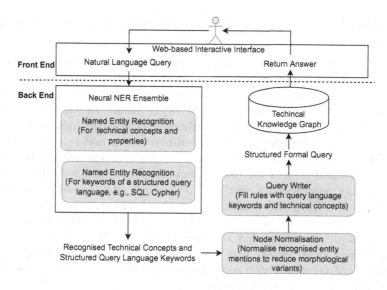

Fig. 1. The model pipeline.

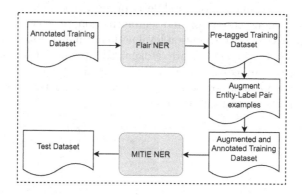

Fig. 2. Named Entity Recognition (NER) Ensemble.

Ensemble Neural Named Entity Recognition. As shown in Fig. 2, the NER Ensemble consists of two different pre-trained NER models. One is Flair Named Entity Recogniser (**Flair NER** [1]). Flair NER incorporates a high quality pre-trained character-level language model for domain specific technical concepts extraction. A character-level language model is chosen because it is good at dealing with inconsistent spelling and spelling error. For example, `valve` and `valv` would have high similarity in word embedding. The other is MITIE Named Entity Recogniser (**MITIE NER**), trained to identify keywords of a structured query language to determine the type of queries. For example, `how many` is mapped to aggregation keyword `count` in SQL and Cypher, in order to trigger different rules. It is trained using the open-source MIT Information Extraction (MITIE) TOOLKIT[8] We specify two different NERs for the pur-

---

[8] MITIE. https://github.com/mit-nlp/MITIE.

pose of identifying graph node entities and structured query language keywords separately.

Node Normalisation. In general, natural language utterances often feature many different forms of the same word. In other words, many word-forms can relate to a single unit of meaning, which is referred as lexeme in linguistics. For example, `replacing`, `replaced`, and `replacement`, or `valves` and `valve`. To reduce inflectional forms and sometimes derivationally related forms of a word to a common base form[9], we use both stemming and lemmatization processes. However, they differ in their flavor. Lemmatization only removes inflectional endings and returns the base or dictionary form of a word (aka lemma). The stemming process chops off the ends of words often including the removal of derivational affixes, and it is a more crude heuristic process. Some domain examples are introduced in the application section. Given that lemmatization and stemming processes recover the base form of a word from its variant differently, some morphological clues are taken advantage of. For example, `-ment` is a suffix that combines with some verbs to produce a noun, e.g. `repair` → `repairment` and `replace`→ `replacement`.

In order to align detected named entities with existing nodes, given a technical KG, a node normalization module is proposed. It includes mention-level part-of-speech tagging, stemming and lemmatization, as well as some morphology clues. Figure 3 shows the mention level node normalization workflow. Tokens in bigram or multi-word mentions are retokenized into one single token before the node normalisation process. For example, ``replaced not working valves'' outputs a list of `replace`, `not working` and `valve`, after the processes of NER Ensemble and node normalisation modules.

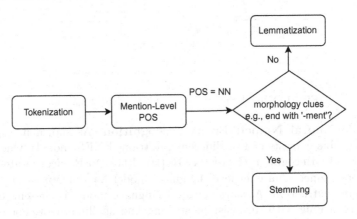

Fig. 3. Node normalisation process.

Rule-Based Structured Formal Query Generator. The generator makes use of the results from NER ensemble and specified technical domain rules.

---

[9] https://nlp.stanford.edu/IR-book/html/htmledition/stemming-and-lemmatization-1.html.

The extracted and aligned named entities and formal query keywords are fillers for a set of structured query language templates, namely `Item-Activity`, `Item-Observation` and `Item-Structured Field`. Manually defined rules are used to generate a valid structured formal query. Take "`show me all leaking items`" as an example, if we observe one of the frequent rules `Item-Observation` w.r.t. the graph pattern `(n1:Node)-[r:HAS_OBSERVATION]->(n2:Node)` in maintenance domain, as long as the head and tail nodes (denoted as n1 and n2) are recognised by our NER ensemble with `Item` and `Observation`, then we are able to generate the corresponding Cypher query (see Sects. 4 and 5 for more details). Once obtained, the corresponding graph engine is able to execute it and return results to the interface.

4 Application

In the context of industrial maintenance, maintenance work orders, stored in a text field, contain rich information about observations and activities that are carried out on assets. Unlocking such information is critical for supporting data-driven decisions for many downstream tasks, such as estimating meantime to failure [5], predictive and prognostic maintenance. It is however expensive and almost impossible to obtain natural and structured query pairs to train neural sequence-to-sequence models despite their state-of-the-art success in neural question answering. In addition, to ensure acceptance by the maintenance engineers, we need explainable models that return precise answers, rather than approximations, to learn by understanding precisely what actions were taken. This makes the hybrid neural and rule-based approach much appealing in this industrial application setting.

4.1 Overview of Maintenance KG

To facilitate this research, we constructed a maintenance KG using text from 500 work orders[10]. This is a tiny fraction of the millions of maintenance work orders collected over years for large resource industries. This small KG is stored in the format of RDF and currently consists of 9,424 triples, including 500 document nodes, 63 FLOC (Functional Location) instances, 252 annotated concept instances using 11 pre-defined entity classes (i.e., `Specifier`, `Time`, `Location`, `Observation`, `Agent`, `Attribute`, `Cardinality`, `Consumable`, `Event`, `Item`, `Activity`) and 586 links using 133 unique hierarchical entity classes from ISO 15926-4[11]. The graph is stored in a Neo4j database, and queried using the Cypher query language. Table 2 shows a comparison of the size of this small KG as compared to benchmark KGs. The statistics of all relations of maintenance knowledge graph is in Table 3. Among the graph nodes, there are document nodes, which contain structured fields (such as `cost`, `start time`). Entities that appear in those documents are linked via a `APPEARS_IN` relationship.

---

[10] `Echidna`: the visualisation of the maintenance KG with form controls to support navigation, is available at https://nlp-tlp.org/maintenance_kg/.

[11] `ISO 15926-4`: https://www.iso.org/standard/73830.html.

Table 2. The statistics of benchmarks and industrial maintenance KG.

Dataset	# Entities	# Relations	# Triple
FB15K	14,951	1,345	483,142
FB15K-237	14,505	237	272,115
WN18	40,943	18	141,442
NELL995	63,361	200	104,384
Maintenance KG	**1,401**	**14**	**9,424**

Table 3. The statistics of maintenance KG relations.

Relation	Count	Relation	Count
APPEARS_IN	3,880	INSTANCE_OF	1,406
HAS_ACTIVITY	1,244	HAS_OBSERVATION	878
HAS_LOCATION	652	APPEARS_WITH	470
SCO	392	HAS_CONSUMABLE	174
HAS_AGENT	130	HAS_SPECIFIER	76
HAS_CARDINALITY	70	HAS_ATTRIBUTE	22
HAS_TIME	22	HAS_EVENT	8

4.2 Neural Named Entity Recognition Ensemble

Flair NER Dataseet. The Flair NER dataset is derived from 3200 work orders (for training set with 20,937 entities), 401 work orders (for development set with 2,652 entities), and 401 work orders (for testing set with 2,702 entities). These work orders were manually annotated by a team of annotators [26]. This domain dataset features 11 entity types in total, capturing a range of maintenance-specific concepts. Then, they are linked to ISO 15926-4 reference data library to obtain structured hierarchical labels in total 111 entity labels including uncategorised.

MITIE NER Dataset. For the purpose of modelling utterance types, two small set of MITIE NER training and testing datasets are used. These datasets cover the three different types of natural language queries. They are list queries, aggregation queries and complex queries (see details in Sect. 5). The average sentence length is 6.2 tokens. Apart from the Flair NER manual annotation entity labels set, MITIE NER is trained with an additional 7 labels (i.e., Match, Count, Avg, Max, Min, FLOC, Cost), where Match is Cypher clause keyword, FLOC and Cost are domain specific technical concepts, and the rest are Cypher aggregation keywords. Manually annotating the training sentences results in 508 training pairs with 18 entity types. Each entity in training pairs is marked as either Cypher keyword, graph node label, or graph node name.

Node Normalisation. As aforementioned, the node normalization process aims to align detected named entities with the given maintenance KG schema. For instance, not and working in a query are re-annotated as one entity mention not working and labeled as Observation. For lemmatization, we normalize named entities to its base dictionary forms, for example, leaking is lemmatized to leak, and valves to valve. For the case that noun words combined verbs with suffix -ment, such as replacement and establishment, their base forms replace and establish respectively will be obtained through stemming process rather than lemmatization process. If stemming is used on valves, the last second letter e will be chopped along with the last letter s. In other words, if the base form of a word ends with e, the stemming process will chop e off as well. Based on our observations, it is better to combine lemmatization and stemming to ensure the results are still dictionary words.

Rule-Based Cypher Query Generator. Typically, there are three graph path patterns appearing most likely in the maintenance KG. We exploit the observed patterns and natural language trigger patterns to generate three rules (see Table 4). The first one is Item-Activity corresponding with its graph path pattern. The second one is Item-Observation taken as the signal to trigger the similar graph path as the first one but with HAS_OBSERVATION binary relationship. The last one is Item-Structured Field referring to the query on structured fields, such as work order actual total cost.

Table 4. The graph path pattern and related rules trigger signals.

Rules	Graph path patterns
Item-Activity	(n1 : Node) − [r : HAS_ACTIVITY]−> (n2 : Node)
Item-Observation	(n1 : Node) − [r : HAS_OBSERVATION]−> (n2 : Node)
Item-Structured Field	(d : Document) <−[: APPEARS_IN] − (n1 : Node) <−[r] − (n2 : Node) − [: APPEARS_IN]−> (d)

5   Results and Discussion

5.1   Ensemble NER Performance Analysis

We evaluate the performances of ensemble NER by computing precision, recall, and F1 scores. Precision is the ratio of actual (or true) positive predictions out of all positive predictions by calculating $TP/(TP + FP)$, where TP refers to true positive and FP is false positive. Recall is the ratio of actual true positive predictions out of all positive examples in the dataset by calculating $TP/(TP + FN)$, where FN denotes false negative. F1 provides a single score that balances both the concerns of precision and recall, namely $F1 = 2 * precision * recall/(precision + recall)$.

Table 5 shows the results of the Flair NER and MITIE NER. In terms of Flair NER, the F1 score of each entity class for MWO concepts is in the range between 0.8 and 0.9, except Attribute and Item. The F1 score of all entity

Table 5. The results of ensemble NER.

Flair NER				MITIE NER			
Entity class	P	R	F1	Entity class	P	R	F1
Activity	0.909	0.905	0.907	Match	1.000	0.933	0.966
Agent	0.784	1.000	0.879	Count	1.000	1.000	1.000
Attribute	0.615	0.667	0.640	Avg	1.000	1.000	1.000
Cardinality	0.938	0.882	0.909	Max	1.000	1.000	1.000
Consumable	0.850	0.810	0.829	Min	1.000	1.000	1.000
Item	0.751	0.817	0.783	FLOC	1.000	1.000	1.000
Location	0.836	0.904	0.868	Cost	0.857	0.750	0.800
Observation	0.810	0.814	0.812				
Specifier	1.000	1.000	1.000				
Time	1.000	0.941	0.970				

classes in MITIE NER surpass 0.8. Currently, we do not evaluate the final system output, due to the absence of an annotated structured query language. The ensemble NER improves the semantic analysis of maintenance work orders in the form of syntactic structures. Misclassified outputs of the ensemble NER module contributes to the construction of candidate structured query language. Further human evaluation may be conducted by annotating Cypher queries w.r.t. their natural language queries.

5.2   Question Types Discussion

Identification of question types is very important to extract relevant information from user questions. When a user posts a query in natural language, our NER ensemble model assigns related question type labels and triggers one of the rules to generate a valid query. A small set of natural language queries is manually created based on three question types. We give a brief explanation of each kind of query types. Table 6 represents several examples and we test their generated Cypher queries by checking the correctness of the returned answers.

List Query. List queries commonly need a list of entities connected with relations in the form of sub-graphs. For instance, `show me all leaking pumps`. One of the problem asked in list type query is fixing the threshold value for quantity of the entity or the number. Our proposed model can handle the list queries with a threshold by being able to map the threshold words to Cypher `Limit` clause. We also create a set of list type query paraphrasing for the model coverage, namely, `show, tell, list, find, return, what`.

Aggregation Query. Aggregation queries compute over all the matching paths. For example, `"how many engines shut down?"` is asked for the total number of

Table 6. The various examples of natural language questions in maintenance domain and their corresponding translated Cypher queries.

	List Query Examples
Input:	Show me all leaking items.
Cypher Query:	`MATCH (n1:Item)-[r1:HAS_OBSERVATION]->(n2:Observation {name: "leak"})` `RETURN n1, r1, n2`
Input:	Show me ten leaking valves.
Cypher Query:	`MATCH (n1:Item {name: "valve" }) -[r1:HAS_OBSERVATION]-> (n2:Observation {name: "leak" })` `RETURN n1, r1, n2 limit 10`
Input:	Show me not working items.
Cypher Query:	`MATCH (n1:Item) -[r1:HAS_OBSERVATION]-> (n2:Observation name: "not working" )` `RETURN n1, r1, n2`
Input:	What assets are required repairment?
Cypher Query:	`MATCH (n1:Item name: "asset" ) -[r1:HAS_ACTIVITY]-> (n2:Activity name: "repair")` `Return n1, r1, n2`
	Aggregation Query Examples
Input:	How many engines shut down?
Cypher Query:	`MATCH (n1:Item {name: "engine" }) -[r1:HAS_OBSERVATION]-> (n2:Observation {name: "shut down" })` `RETURN COUNT(*)`
Input:	What is the maximum number of leaking pumps?

Cypher Query:	`MATCH (n1:Item {name: "pump" }) -[r1:HAS_OBSERVATION]-> (n2:Observation {name: "leak" })`
	`WITH n1, n2, COUNT(n1) as total`
	`RETURN MAX(total) as maximumNumber`

Input:	Tell me the average number of leaking items?

Cypher Query:	`MATCH (n1:Item)-[r1:HAS_OBSERVATION]->(n2:Observation {name: "leak"})`
	`WITH n1, n2, COUNT(n1) as total`
	`RETURN AVG(n1)`

	Complex Query Examples
Input:	Show me the top three recently not working items.

Cypher Query:	`MATCH (n1:Item) -[r1:HAS_OBSERVATION]-> (n2:Observation {name: "not working" })`
	`WITH n1, n2, COUNT(n1) as total`
	`ORDER BY total DESC`
	`RETURN collect(n1)[0] as malFunctionalItem, n2 limit 3`

Input:	Show me the total number of rotating equipment whose cost is between 2000 and 50100.

Cypher Query:	`MATCH (d:Document)<-[a:APPEARS_IN]-(n1: 'Item/Rotating_Equipment')`
	`<-[r]-(n2:'Item/Rotating_Equipment')-[:APPEARS_IN]->(d)`
	`WHERE d.Work_Order_Total_Actual_Cost >= 2000 AND d.Work_Order_Total_Actual_Cost <= 50100`
	`WITH n1, n2, COUNT(n1) as total1, COUNT(n2) as total2`
	`RETURN SUM(total1, total2)`

shutting down engines. One of the advantages of aggregation queries in the web-based interface is some users may like statistical information to have a general understanding of data, including counting, maximum, minimum and average.

Complex Query. complex query examples are more difficult to parse. Questions such as "show me the total number of rotating equipment whose cost is between 2000 and 50100" normally requires synthesizing information from multiple documents to get answers. It requires the entity type information of rotating equipment, the structured field of functional location range and the aggregation type of total number of the nodes constrained by the aforementioned information. The benefit of complex type queries is to save the multi-steps queries via drop-down lists into one step and makes the natural language queries more flexible than structured check boxes queries.

6   Conclusions and Future Work

This paper presents a hybrid neural/rule-based approach to constructing formal queries for knowledge graphs in low-resource technical domains. We demonstrate the application of using the proposed pipeline to support natural language queries for an industry maintenance KG. The proposed model provides three benefits: 1) support queries which are not able to be realised via pre-defined structured navigation, 2) reduce multiple steps queries into one step, 3) support some of the complex queries.

Currently the diversity of our training dataset is limited and our model does not incorporate the graph schema. There is no guarantee that the user is familiar with the data and schema. Therefore, the model should not only be able to build a valid formal query with respect to the user's input, but should also have the generalization capability of being robust to variations in the way each user expresses a query in natural language. In the future, we will create our large-scale question and Cypher query pairs as our training dataset and advance our model architecture to take into account the graph schema along with natural language utterances. Another possible direction is to bridge the gap between two different domain fields, for example biomedical field and maintenance domain, by solving the common case in a domain-independent manner.

Acknowledgment. This research is supported by the Australian Research Council through the Centre for Transforming Maintenance through Data Science (grant number IC180100030), funded by the Australian Government. We thank the reviewers for their insightful comments, and Tyler Bikaun for his proofreading and suggestions.

References

1. Akbik, A., Bergmann, T., Blythe, D., Rasul, K., Schweter, S., Vollgraf, R.: Flair: an easy-to-use framework for state-of-the-art NLP. In: Proceedings of the 2019 Conference of the North American Chapter of the Association for Computational Linguistics (Demonstrations), pp. 54–59 (2019)
2. Auer, S., Bizer, C., Kobilarov, G., Lehmann, J., Cyganiak, R., Ives, Z.: DBpedia: a nucleus for a web of open data. In: Aberer, K., et al. (eds.) ASWC/ISWC -2007. LNCS, vol. 4825, pp. 722–735. Springer, Heidelberg (2007). https://doi.org/10.1007/978-3-540-76298-0_52
3. Bahdanau, D., Cho, K., Bengio, Y.: Neural machine translation by jointly learning to align and translate. arXiv preprint arXiv:1409.0473 (2014)
4. Berant, J., Chou, A., Frostig, R., Liang, P.: Semantic parsing on freebase from question-answer pairs. In: Proceedings of the 2013 Conference on Empirical Methods in Natural Language Processing, pp. 1533–1544 (2013)

5. Bikaun, T., Hodkiewicz, M.: Semi-automated estimation of reliability measures from maintenance work order records. In: PHM Society European Conference, vol. 6, p. 9 (2021)

6. Bollacker, K., Evans, C., Paritosh, P., Sturge, T., Taylor, J.: Freebase: a collaboratively created graph database for structuring human knowledge. In: Proceedings of the 2008 ACM SIGMOD International Conference on Management of Data, pp. 1247–1250 (2008)

7. Chakraborty, N., Lukovnikov, D., Maheshwari, G., Trivedi, P., Lehmann, J., Fischer, A.: Introduction to neural network based approaches for question answering over knowledge graphs. arXiv preprint arXiv:1907.09361 (2019)

8. Dahl, D.A., et al.: Expanding the scope of the ATIS task: the ATIS-3 corpus. In: Human Language Technology: Proceedings of a Workshop held at Plainsboro, New Jersey, 8–11 March 1994 (1994)

9. Devlin, J., Chang, M.W., Lee, K., Toutanova, K.: BERT: pre-training of deep bidirectional transformers for language understanding. arXiv preprint arXiv:1810.04805 (2018)

10. Dong, L., Lapata, M.: Language to logical form with neural attention. arXiv preprint arXiv:1601.01280 (2016)

11. Ferrucci, D., et al.: Building Watson: an overview of the DeepQA project. AI Mag. **31**(3), 59–79 (2010)

12. Francis, N., et al.: Cypher: an evolving query language for property graphs. In: Proceedings of the 2018 International Conference on Management of Data, pp. 1433–1445 (2018)

13. Grishman, R., Sundheim, B.: Message understanding conference-6: a brief history. In: COLING 1996 Volume 1: The 16th International Conference on Computational Linguistics (1996). https://aclanthology.org/C96-1079

14. Hendrix, G.G., Sacerdoti, E.D., Sagalowicz, D., Slocum, J.: Developing a natural language interface to complex data. ACM Trans. Database Syst. (TODS) **3**(2), 105–147 (1978)

15. Hochreiter, S., Schmidhuber, J.: Long short-term memory. Neural Comput. **9**(8), 1735–1780 (1997)

16. Jia, R., Liang, P.: Data recombination for neural semantic parsing. arXiv preprint arXiv:1606.03622 (2016)

17. Kirsch, R.A.: Computer interpretation of English text and picture patterns. IEEE Trans. Electron. Comput. **4**, 363–376 (1964)

18. LeCun, Y., Bengio, Y., et al.: Convolutional networks for images, speech, and time series. Handb. Brain Theory Neural Netw. **3361**(10), 1995 (1995)

19. Lin, X.V., Socher, R., Xiong, C.: Bridging textual and tabular data for cross-domain text-to-SQL semantic parsing. arXiv preprint arXiv:2012.12627 (2020)

20. Pennington, J., Socher, R., Manning, C.: GloVe: global vectors for word representation. In: Proceedings of the 2014 Conference on Empirical Methods in Natural Language Processing (EMNLP), Doha, Qatar, pp. 1532–1543. Association for Computational Linguistics (2014). https://doi.org/10.3115/v1/D14-1162. https://aclanthology.org/D14-1162

21. Peters, M.E., et al.: Deep contextualized word representations. In: Proceedings of the 2018 Conference of the North American Chapter of the Association for Computational Linguistics: Human Language Technologies, New Orleans, Louisiana, Volume 1 (Long Papers), pp. 2227–2237. Association for Computational Linguistics (2018). https://doi.org/10.18653/v1/N18-1202. https://aclanthology.org/N18-1202

22. Pourhabibi, T., Ong, K.L., Kam, B.H., Boo, Y.L.: Fraud detection: a systematic literature review of graph-based anomaly detection approaches. Decis. Support Syst. **133**, 113303 (2020)
23. Rumelhart, D.E., Hinton, G.E., Williams, R.J.: Learning internal representations by error propagation. Technical report. California Univ San Diego La Jolla Inst for Cognitive Science (1985)
24. Scarselli, F., Gori, M., Tsoi, A.C., Hagenbuchner, M., Monfardini, G.: The graph neural network model. IEEE Trans. Neural Netw. **20**(1), 61–80 (2008)
25. Sorokin, D.: Knowledge graphs and graph neural networks for semantic parsing (2021)
26. Stewart, M., Hodkiewicz, M., Liu, W., French, T.: MWO2KG and Echidna: constructing and exploring knowledge graphs from maintenance data. In: Proceedings of the Institution of Mechanical Engineers, Part O: Journal of Risk and Reliability (in press)
27. Suchanek, F.M., Kasneci, G., Weikum, G.: YAGO: a core of semantic knowledge. In: Proceedings of the 16th International Conference on World Wide Web, pp. 697–706 (2007)
28. Sun, C., et al.: A natural language interface for querying graph databases. Master thesis, Massachusetts Institute of Technology (2018)
29. Tang, L.R., Mooney, R.J.: Using multiple clause constructors in inductive logic programming for semantic parsing. In: De Raedt, L., Flach, P. (eds.) ECML 2001. LNCS (LNAI), vol. 2167, pp. 466–477. Springer, Heidelberg (2001). https://doi.org/10.1007/3-540-44795-4_40
30. Wang, X., He, X., Cao, Y., Liu, M., Chua, T.S.: KGAT: knowledge graph attention network for recommendation. In: Proceedings of the 25th ACM SIGKDD International Conference on Knowledge Discovery & Data Mining, pp. 950–958 (2019)
31. Weizenbaum, J.: Eliza—a computer program for the study of natural language communication between man and machine. Commun. ACM **9**(1), 36–45 (1966)
32. Wilks, Y., Fass, D.: The preference semantics family. Comput. Math. Appl. **23**(2–5), 205–221 (1992)
33. Woods, W.A.: Progress in natural language understanding: an application to lunar geology. In: Proceedings of the June 4–8, 1973, National Computer Conference and Exposition, pp. 441–450 (1973)
34. Xu, K., Reddy, S., Feng, Y., Huang, S., Zhao, D.: Question answering on freebase via relation extraction and textual evidence. In: Proceedings of the 54th Annual Meeting of the Association for Computational Linguistics (Volume 1: Long Papers) (2016)
35. Yadav, V., Bethard, S.: A survey on recent advances in named entity recognition from deep learning models. arXiv preprint arXiv:1910.11470 (2019)
36. Yih, S.W., Chang, M.W., He, X., Gao, J.: Semantic parsing via staged query graph generation: question answering with knowledge base. In: Proceedings of the Joint Conference of the 53rd Annual Meeting of the ACL and the 7th International Joint Conference on Natural Language Processing of the AFNLP (2015)
37. Yih, W., Richardson, M., Meek, C., Chang, M.W., Suh, J.: The value of semantic parse labeling for knowledge base question answering. In: Proceedings of the 54th Annual Meeting of the Association for Computational Linguistics (Volume 2: Short Papers), pp. 201–206 (2016)
38. Yu, T., et al.: CoSQL: a conversational text-to-SQL challenge towards cross-domain natural language interfaces to databases. arXiv preprint arXiv:1909.05378 (2019)
39. Yu, T., et al.: Spider: a large-scale human-labeled dataset for complex and cross-domain semantic parsing and text-to-SQL task (2018)

40. Zelle, J.M., Mooney, R.J.: Learning to parse database queries using inductive logic programming. In: AAAI/IAAI, vol. 2 (1996)
41. Zettlemoyer, L.S., Collins, M.: Learning to map sentences to logical form: structured classification with probabilistic categorial grammars. arXiv preprint arXiv:1207.1420 (2012)
42. Zheng, W., Cheng, H., Zou, L., Yu, J.X., Zhao, K.: Natural language question/answering: let users talk with the knowledge graph. In: Proceedings of the 2017 ACM on Conference on Information and Knowledge Management, pp. 217–226 (2017)
43. Zhong, V., Xiong, C., Socher, R.: Seq2SQL: generating structured queries from natural language using reinforcement learning. CoRR abs/1709.00103 (2017)

Decomposition of Service Level Encoding for Anomaly Detection

Rob Muspratt[1,2]([⊠]) [iD] and Musa Mammadov[1] [iD]

[1] School of Information Technology, Deakin University, Melbourne, Australia
{rmuspratt,musa.mammadov}@deakin.edu.au
[2] Transport Accident Commission, Geelong, Australia

Abstract. Application of anomaly thresholds to health provider service level data encoded in vector form can lead to unintended consequences in terms of output complexity and business appropriate interpretation. In this paper we show that specific analysis incorporating feature decomposition with the application of relative business knowledge prior to selecting anomaly thresholds is an effective strategy across multiple health provider disciplines for addressing this complexity. Cluster definitions of Modal, Specialised and Aberrant are introduced as a descriptive framework on which to interpret said feature decomposition and to aid threshold setting. This strategy furthers the work introduced in [1] as refinement to an existing anomaly detection scheme.

Keywords: Healthcare provider · Anomaly detection · Feature decomposition · Outlier thresholds

1 Introduction

The Transport Accident Commission of Victoria (TAC) is a State Government owned organisation whose key function is funding treatment and support services for people injured in transport accidents. The TAC generates health provider billing transactions as a by-product of processing and funding health provider accounts and services for its clients. It is the descriptive attributes of these transactions which constitute the output variables of this study (e.g. service item selection, amount paid, date of service, etc.).

Use of the term anomaly and the range of alternate descriptors available depends highly on the domain of application, e.g. outliers, discordant observations, exceptions, aberrations, surprises, peculiarities or contaminants [2]. Whilst previous research upon which this investigation is based was concerned with defined outliers, it is more appropriate to apply the anomaly descriptor in this instance. The term "anomaly" more accurately reflecting the "non-extreme" nature of discordant observations sought [3].

Consideration of appropriate output translation and targeting of provider-client combinations along with computational requirements and reusability led to the development of a bespoke method based on direct provider comparison [1]. In terms of actions or behaviours over similar cases/claims the method utilises the following scheme:

$$\text{Input} \rightarrow \text{Provider} \rightarrow \text{Output}$$

L. A. F. Park et al. (Eds.): AusDM 2022, CCIS 1741, pp. 192–204, 2022.
https://doi.org/10.1007/978-981-19-8746-5_14

Input is assumed to combine a set of features that could be used to define "similar clients", for example, age, gender, postcode, injury types, etc. Output is assumed to be a set of responses/actions by a particular provider, for example this may include service types, service levels, service intensities, billed amount, type of billing, etc.. In this paper we develop further the approach introduced in [1] by considering and addressing the following two important issues with regard to encoded service levels:

- Unusual patterns are evident in the encoding of service levels which require additional investigation, decomposition and analysis to ensure relevance, from a business perspective, of any outputs.
- Applying a best-fit distribution to each component makes clear the divergence that exists in the tail of a number of the sub-groups and highlights the presence of non-extreme anomalies of interest.

As with any data mining exercise, feature generation and selection are critical to the final results [4]. By addressing the two issues above we will show that effective feature generation of an important business characteristic (service level selection) coupled with appropriate decomposition and analysis leads to valuable domain insights. The results obtained in Sect. 3 show a consistent theme identified between health disciplines investigated in this paper. This theme is the existence of 3 sub-groups or clusters in the service level population which have been titled Modal, Specialised and Aberrant. All 3 clusters require specific analysis via decomposition to effect appropriate anomaly identification.

2 Algorithm and Notations

Distance functions for input (x) and output (y) variables will be denoted by d^x and d^y respectively.

2.1 Input/Output Spaces

Consistent with previous analysis [1] the following input variables are used to select similar claims among all sample data points:

I1: List of Injuries - List of all 20 possible client injuries. The resulting Injury Vector $I = (I_1, \ldots, I_{20})$ is a binary vector of length 20 representing corresponding incidence of injury. Depending on the severity of injuries subsequent weighting are applied in the form

$$I_k = 100 \, I_k / k^2, \; k = 1, \ldots, 20.$$

I2: Variable "Age", denoted by A, has values in the interval $(0, 100)$, these values are rescaled to the range $[0, 5]$.

I3: Variable "Time from Accident", denoted by T, is also rescaled to the range $[0, 5]$.

After scaling all the input variables $x = (I, A, T)$ the Euclidean distance would be the best choice to define a neighborhood in the input space around a given data point in D, defined as the distance function d^x.

The output variable considered in this paper is the following:

O1: Service Levels - A vector of services $s = (s_1, ..., s_L)$ defined for each provider-claim combination where L is the number of service levels, and s_l is the number of services of level $l \in \{1, ..., L\}$.

When considering Service Levels, measuring the proportions of billing at each level would be of most interest, accordingly a Cosine measure is best in this case. Therefore, distance d^y between two services s^1 and s^2 will be defined in the form:

$$d^y(s^1, s^2) = 1 - Cos(s^1, s^2)$$

2.2 Algorithm

The algorithm used in this paper is described by Steps 1–3 below, where Step 1 is adopted from [1].

Step 1: Given a health provider-claim (or patient) combination (p, c), the degree of abnormality with regard to clinical treatment billed is calculated thus. Consider an arbitrary data point $(p^0, c^0; x^0, y^0)$ in our domain, D.

- Calculate the distance $d^x(x^0, x)$ from all data points $(p, c; x, y) \in D$ and select the closest n^{top} points, the neighbourhood, that will be denoted by N^0.
- Calculate the average value AvS^1 of distance $d^y(y^0, y)$ over all data points in N^0.
- The resultant outlying value $VLM(p^0, c^0) = AvS^1$

Note: This resultant value defines anomalies in terms of the "local" neighbourhood, that is, the divergence with respect to the closest n^{top} claims; we will call it *Variation from Local Mean* (VLM).

Step 2 (Best Fit): Find the best fit distribution function for the outlying value VLM over data D.

Step 3: Define and select appropriate outlier values/intervals for provider service levels.

Finding the best threshold for extreme outlying values on the right tale of the distribution function is an important but difficult problem. The best fit found in **Step 2** is used in this step by analysing the divergence between the best fit and related variable (VLM). It reveals two important points:

- There is often a clear threshold point at the right tale where this divergence occurs.
- There are two intervals on the right tale after the threshold value, namely Sub Extreme (SE) and High Extreme (HE).

2.3 Defining Interval Sub Extreme (SE)

The main rationale in considering the top two subsets SE and HE is as follows. Practice shows that the highest ranked anomalies usually have solid reasons for their large outlying values (as exceptional cases/errors). Accordingly, the most interesting anomalies could be expected to be in the range of SE rather than in HE. Application of a best-fit distribution to each variable reveals any clear divergence that exists in the right tail and will highlight the presence of sub-extreme anomalies of interest.

3 Results

In the numerical experiments that follow, to find the best fit (**Step 2**) we use the scipy.stats package (Python) considering 89 different distribution functions. Optimal parameter estimations are performed in terms of Maximum Likelihood Estimation (MLE) where in all cases the bin size was set to be 200.

3.1 Physiotherapy Service Levels

The first sample dataset used in this study contains aggregated Physiotherapy billing data of 31,447 health provider/client combinations and is representative of 396,472 underlying transactions over a 30 month period. Local neighbourhood in this instance has been set at the closest 100 points. For simplicity of labelling local distance observations are referred to as the variation from local mean (VLM).

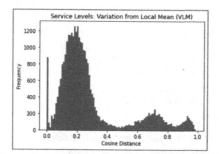

Fig. 1. Histogram for Physiotherapy Service Levels (Cosine distance measure) applying Variation from Local Mean (VLM) as defined. Initial inspection show an assumption of normality is inconsistent with observed results. Service Levels applying VLM requires further decomposition and investigation to understand its component features.

Physiotherapy Service Levels are clearly a combination of distinct sub populations noted by the multiple peaks in the related VLM histogram (Fig. 1). This is attributed to both specific service level preferences within provider sub-groups (e.g. Neurophysio extended head/spinal injury consultations or less expensive hydrotherapy/group sessions) and the nature of the initial encoding of these service levels (i.e. equivalent time based encoding of consultations <30 min, =30 min and >30 min in duration). Service levels correspond to this time based encoding and are referred to as service level 1, 2 and 3 respectively. For simplicity in data handling and analysis the service level combinations are represented by a binary vector corresponding to the use of each of the 3 service levels for a particular provider-client combination. With 3 service levels this gives a maximum of 7 valid combinations represented by the binary vectors 001, 010, 011, 100, 101, 110 and 111 (e.g. 001 represents service level 3 only, 010 represents service level 2 only, etc.).

To define appropriate local anomaly thresholds it is necessary to examine service level combinations in separate but related groupings. Given service level 2 (=30 min) is the most prevalent service modality, all service levels containing level 2 are considered together (i.e. 010, 011, 110 and 111). Divergence from service level 2 can be an indicator of over-servicing by a provider and warrants further investigation or review (Fig. 2).

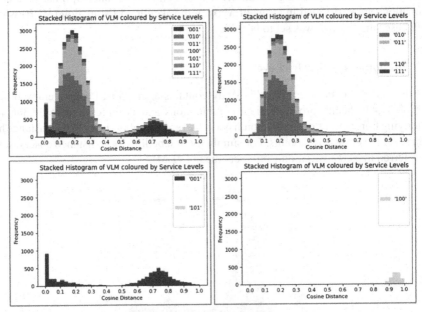

Fig. 2. Stacked Histograms for Physiotherapy Service Levels (Cosine distance measure) showing overall distribution along with decomposition into Modality Servicing (010, 011, 110 and 111), Neurophysio Treatment (001 and 101) and Group Treatment (100) using Variation from Local Mean (VLM).

Modality servicing (categories 010, 011, 110 and 111) does not exhibit normality and hence original z-value thresholds applied in [1] corresponding to the top 3% and 5% of the population (1.4 and 1.88 respectively) are not appropriate. A mielke distribution suggests divergence from the tail between cosine distance values of 0.52 and 0.7 as the best fit alterative (Fig. 3).

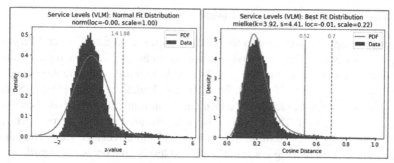

Fig. 3. Histograms for Physiotherapy Service Levels (Cosine distance measure) showing Modality Servicing (010, 011, 110 and 111) with normal and best fit distributions of VLM (PDF = Probability Density Function of relevant distribution). Minor tail divergence evident and original z-value thresholds corresponding to the top 3% and 5% of the population (1.4 and 1.88 respectively) are replaced by cosine distance values of 0.52 and 0.7.

Likewise Service Levels 001 and 101 with VLM $\geq$ 0.45 have more appropriate thresholds of divergence identified between cosine distance 0.86 and 0.92 when applying a minimally transformed normal distribution. The extreme nature and minimal business value of the remaining service level group, 100, is cause for its omission from further VLM anomaly calculations in this case (Fig. 4).

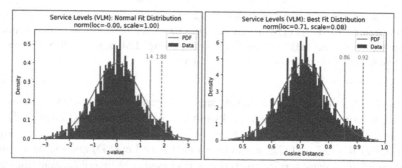

Fig. 4. Histograms for Physiotherapy Service Levels (Cosine distance measure) showing Service Levels 001 and 101, VLM $\geq$ 0.45, with normal and best fit distributions of VLM (PDF = Probability Density Function of relevant distribution). Prominent tail divergence evident and original z-value thresholds corresponding to the top 3% and 5% of the population (1.4 and 1.88 respectively) are replaced by cosine distance values of 0.86 and 0.92.

It is clear that use of the cosine distance measure with this input domain leads to an abnormal distribution at the local level requiring appropriate handling. More specifically there are two local subgroups which benefit from a tailored threshold, the modal and the 001/101 service levels with VLM $\geq$ 0.45.

3.2 General Practitioner Service Levels

The second sample dataset used in this study contains aggregated General Practitioner billing data of 35,116 health provider/client combinations and is representative of 215,045 underlying transactions over a 24 month period. Local neighbourhood again has been set at the closest 100 points. The Medicare Benefit Schedule [5] (MBS) defines the majority of General Practitioner consultations in four categories based on duration (Levels A, B, C and D). These categories, or levels, are prevalent across face-to-face consultations in a clinical setting, home visits and more recently via telehealth (telephone or video conferencing consultations). The levels are consistent regardless of the delivery type and correspond to the following: Level A – less than 6 min; Level B – 6 to 20 min; Level C – 21 to 40 min; Level D – greater than 40 min in duration.

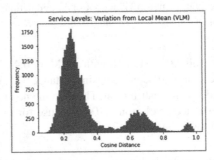

Fig. 5. Histogram for General Practitioner Service Levels (Cosine distance measure) applying Variation from Local Mean (VLM) as defined. Initial inspection show an assumption of normality is inconsistent with observed results. Service Levels applying VLM requires further decomposition and investigation to understand its component features.

General Practitioner Service Levels are also clearly a combination of distinct sub populations noted by the multiple peaks in the related VLM histogram (Fig. 5). This is attributed to both specific service level preferences amongst providers (e.g. frequent use of Level B and Level C consults over Level A or Level D) and the nature of the initial encoding of these service levels (i.e. MBS based encoding of consultations based on duration). Service levels correspond to this time based encoding and are referred to as service level 1, 2, 3 and 4 respectively. With 4 service levels the binary vector representation gives a maximum of 15 valid combinations represented by the binary vectors 0001, 0010, ..., 1111 (e.g. 0001 represents service level D only, 0010 represents service level C only, etc.).

Defining appropriate local anomaly thresholds again requires examining service level combinations in separate but related groupings. Service level B (0100) is the most prevalent service modality, hence all service levels containing level B are considered together along with the majority of level A (i.e. 0100, 0101, 0110, 0111, 1001, 1010, 1011, 1100, 1101, 1110 and 1111) (Fig. 6).

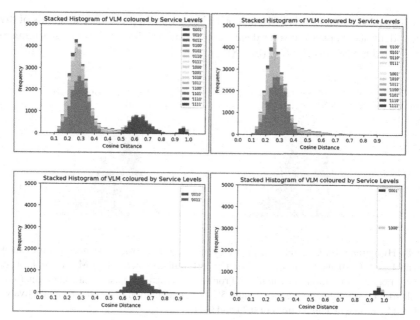

Fig. 6. Stacked Histograms for General Practitioner Service Levels (Cosine distance measure) showing overall distribution along with decomposition into Modality Servicing (0100, 0101, 0110, 0111, 1001, 1010, 1011, 1100, 1101, 1110 and 1111), Level C (0010 and 0011) and Level D with Level A only (0001 and 1000) using Variation from Local Mean (VLM).

Modality servicing does not exhibit normality and hence original z-value thresholds applied in [1] corresponding to the top 3% and 5% of the population (1.4 and 1.88 respectively) are not appropriate. A mielke distribution suggests divergence from the tail between cosine distance values of 0.47 and 0.59 as the best fit alterative (Fig. 7).

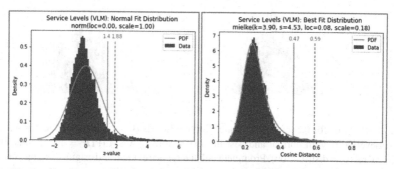

Fig. 7. Histograms for General Practitioner Service Levels (Cosine distance measure) showing Modality Servicing with normal and best fit distributions of VLM (PDF = Probability Density Function of relevant distribution). Minor tail divergence evident and original z-value thresholds corresponding to the top 3% and 5% of the population (1.4 and 1.88 respectively) are replaced by cosine distance values of 0.47 and 0.59.

Likewise Service Levels 0010 and 0011 have more appropriate thresholds of divergence identified between cosine distance 0.79 and 0.84 when applying a minimally transformed normal distribution (Fig. 8).

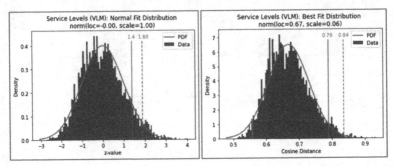

Fig. 8. Histograms for General Practitioner Service Levels (Cosine distance measure) showing Service Levels 0010 and 0011 with normal and best fit distributions of VLM (PDF = Probability Density Function of relevant distribution). Prominent tail divergence evident and original z-value thresholds corresponding to the top 3% and 5% of the population (1.4 and 1.88 respectively) are replaced by cosine distance values of 0.79 and 0.84.

Service Level 0001 on its own has more appropriate thresholds of divergence identified between cosine distance 0.976 and 0.985 when applying a power-normal distribution. The extreme nature and minimal business value of the remaining service level group, 1000, is cause for its omission from further VLM anomaly calculations as was the experience with Physiotherapy (Fig. 9).

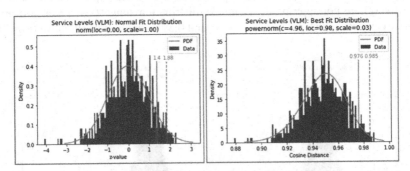

Fig. 9. Histograms for General Practitioner Service Levels (Cosine distance measure) showing Service Level 0001 with normal and best fit distributions of VLM (PDF = Probability Density Function of relevant distribution). Prominent tail divergence evident and original z-value thresholds corresponding to the top 3% and 5% of the population (1.4 and 1.88 respectively) are replaced by cosine distance values of 0.976 and 0.985.

It is clear that use of the cosine distance measure with this input domain also leads to an abnormal distribution at the local level requiring appropriate handling. Most interesting is that the number of service levels does not necessarily translate directly to the number of population sub-groups requiring specific analysis.

3.3 Psychiatric Service Levels

The third sample dataset used in this study contains aggregated Psychiatrist billing data of 3,363 health provider/client combinations and is representative of 34,087 underlying transactions over a 24 month period. Local neighbourhood again has been set at the closest 100 points. The MBS defines the majority of Psychiatric consultations in five categories based on duration (Levels A, B, C, D and E). These categories, or levels, are prevalent across face-to-face consultations in a clinical setting and more recently via telehealth. The levels are consistent regardless of the delivery method and correspond to the following: Level A – less than 15 min; Level B – 15 to 30 min; Level C – 30 to 45 min; Level D – 45 to 75 min; Level E – greater than 75 min in duration (Fig. 10).

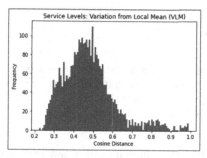

Fig. 10. Histogram for Psychiatric Service Levels (Cosine distance measure) applying Variation from Local Mean (VLM) as defined. Initial inspection show an assumption of normality is inconsistent with observed results. Service Levels applying VLM requires further decomposition and investigation to understand its component features.

Psychiatric Service Levels are also a combination of distinct sub populations which becomes evident upon inspection of the stacked VLM histogram (Fig. 11). This is attributed to both specific service level preferences amongst providers (e.g. frequent use of Level B, C and D consults over Level A or Level E) and the time based nature of the initial encoding. Service levels corresponding to this time based encoding and are referred to as service level 1, 2, 3, 4 and 5 respectively. With 5 service levels the binary vector representation gives 24 valid combinations since 7 service level combinations are not present in the population.

Defining appropriate local anomaly thresholds again requires examining service level combinations in separate but related groupings. Service level D (00010) is the most prevalent service modality followed by the combination of levels C & D (00110) and then level C only (00100). The modality service grouping remains after removing service levels 10000, 01000, 11000 and 00001.

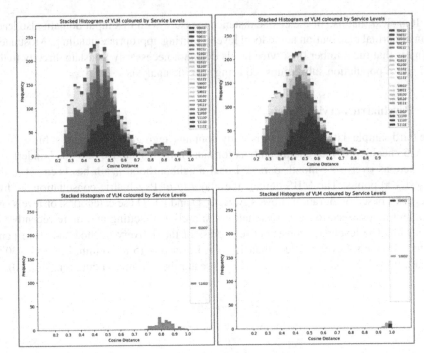

Fig. 11. Stacked Histograms for Psychiatric Service Levels (Cosine distance measure) showing overall distribution along with decomposition into Modality Servicing (predominantly Levels C and D), Level B with Level A/B (01000 and 11000) and Levels A/E only (10000 and 00001) using Variation from Local Mean (VLM).

Modality servicing does not exhibit normality and hence z-value thresholds applied in prior disciplines (1.4 and 1.88 respectively) are not appropriate. A burr distribution suggests divergence from the tail between cosine distance values of 0.69 and 0.79 as the best fit alterative (Fig. 12).

Service levels 01000 and 11000 have more appropriate thresholds of divergence identified between cosine distance 0.895 and 0.935 when applying a power-normal distribution. The extreme nature and minimal business value of the remaining service level group, 10000 and 00001, is cause for its omission from further VLM anomaly calculations.

3.4 Discipline Comparison

Most interesting is that the number of service levels does not necessarily translate directly to the number of population sub-groups requiring appropriate handling. Across all three disciplines of Physiotherapy, General Practice and Psychiatry there exists a modal group or cluster (Modal), a subsequent group comprising specialised treatment options or combinations (Specialised) and an extreme outlier group (Aberrant) that possesses minimal business value by volume and servicing characteristics. Proportions of these 3 clusters for each health discipline are presented in Table 1.

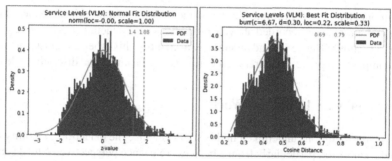

Fig. 12. Histograms for Psychiatric Service Levels (Cosine distance measure) showing Modality Servicing with normal and best fit distributions of VLM (PDF = Probability Density Function of relevant distribution). Original z-value thresholds corresponding to the top 3% and 5% of the population (1.4 and 1.88 respectively) are replaced by cosine distance values of 0.69 and 0.79 (Fig. 13).

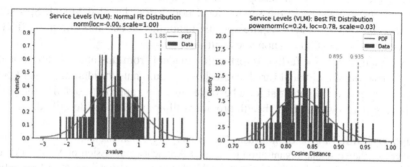

Fig. 13. Histograms for Psychiatric Service Levels (Cosine distance measure) showing Service Levels 01000 and 11000 with normal and best fit distributions of VLM (PDF = Probability Density Function of relevant distribution). Original z-value thresholds corresponding to the top 3% and 5% of the population (1.4 and 1.88 respectively) are replaced by cosine distance values of 0.895 and 0.935.

Table 1. Proportion of provider-client observations evident across each health discipline and service level cluster.

Discipline	Cluster/Sub-group		
	Modal	Specialised	Aberrant
Physiotherapy	77.5%	19.1%	3.4%
General practice	81.6%	16.1%	2.3%
Psychiatry	95.5%	3.7%	0.8%

Without analysis via decomposition and the overlay of appropriate business knowledge all Aberrant cases recorded above would otherwise be included as extreme anomalous observations. Hence it is advantageous to identify this Aberrant cluster as part of the decomposition process and treat it according to its business and discipline relevance.

4   Summary and Further Work

Detailed considerations undertaken in this study enhance the detection capability of the original framework, particularly from a business relevance perspective. Whilst standard threshold setting was considered appropriate initially, it is only upon close inspection and analysis of the service level output variable that appropriate limits and rules can be enacted to increase overall effectiveness of the model. Variable decomposition and threshold setting based on points of divergence from a normal or best-fit theoretical distribution, where appropriate, underpin this outcome. A consistent theme identified between the health provider disciplines investigated in this paper is the existence of 3 sub-groups or clusters in the service level population, labelled Modal, Specialised and Aberrant. All 3 clusters require specific handling via decomposition to effect the appropriate identification of anomalies within.

Two main topics for further investigation evident from this analysis are: potential for optimisation of individual variable thresholds to automate the identification of points of divergence; and the potential for a bespoke clustering approach to define the service level sub-groups evident in the health provider billing populations described.

Compliance with Ethical Standards. The primary/corresponding author is a substantive employee of the custodian organisation (Transport Accident Commission of Victoria) in addition to being a research degree student at Deakin University. The authors have no additional interests to declare that are relevant to the content of this article. Data has been approved for use in a research capacity by the TAC and is fully de-identified to protect individual and health provider privacy. This study has received Deakin University ethics approval (reference number: 2021-363).

References

1. Mammadov, M., Muspratt, R., Ugon, J.: Detection of outlier behaviour amongst health/medical providers servicing TAC clients. In: Boo, Y.L., Stirling, D., Chi, L., Liu, L., Ong, K.-L., Williams, G. (eds.) AusDM 2017. CCIS, vol. 845, pp. 161–172. Springer, Singapore (2018). https://doi.org/10.1007/978-981-13-0292-3_10
2. Chandola, V., Banerjee, A., Kumar, V.: Anomaly detection: a survey. ACM Comput. Surv. **41**(3), 1–58 (2009)
3. Kirlidog, M., Asuk, C.: A fraud detection approach with data mining in health insurance. Procedia Soc. Behav. Sci. **62**, 989–994 (2012)
4. Bahnsen, C., Aouada, D., Stojanovic, A., Ottersten, B.: Feature engineering strategies for credit card fraud detection. Expert Syst. Appl. **51**, 134–142 (2016)
5. Medicare Benefits Schedule. Australian Government Department of Health (2021)

Improving Ads-Profitability Using Traffic-Fingerprints

Adam Gabriel Dobrakowski[1], Andrzej Pacuk[1], Piotr Sankowski[1,2,3(✉)],
Marcin Mucha[3], and Paweł Brach[4]

[1] MIM Solutions, Warsaw, Poland
{adam.dobrakowski,andrzej.pacuk,piotr.sankowski}@mim.ai
[2] IDEAS NCBR, Warsaw, Poland
[3] University of Warsaw, Warsaw, Poland
mucha@mimuw.edu.pl
[4] HitDuck, Warsaw, Poland
pbrach@hitduck.com

Abstract. This paper introduces the concept of *traffic-fingerprints*, i.e., normalized 24-dimensional vectors representing a distribution of daily traffic on a web page. Using specially tuned k-means clustering we show that similarity of traffic-fingerprints is related to the similarity of profitability time patterns for ads shown on these pages. In other words, these fingerprints are correlated with the conversions rates, thus allowing us to argue about conversion rates on pages with negligible traffic. By blocking or unblocking whole clusters of pages we were able to increase the revenue of online campaigns by more than 50%.

Keywords: Ad networks · Traffic-fingerprint · Clustering · k-means

1 Introduction

Internet becomes more and more anonymous which poses new challenges in front of advertisers. On the positive side, we are currently on the way to guarantee users the comfort of not being tracked all the time, e.g., via cookies. On the negative side, as we cannot identify users anymore, it is becoming harder for the advertisers to make sure that a user is interested in the content of an ad. The shift in anonymization paradigm is reaching even further than just users. Nowadays, from the point of view of the advertisers, not only users are being represented by unidentifiable numbers, but also web pages or even ad placements. In particular, this is the case for many advertising platforms connecting advertisers to web pages that provide slots for ads. The key function of advertising networks is an aggregation of ad supply from many different web publishers and matching it

---

This work was supported by the National Centre for Research and Development (NBBR) grant no. POIR.01.01.01-00-0945/19, the National Science Center (NCN) grant no. 2020/37/B/ST6/04179 and the ERC CoG grant TUgbOAT no 772346.

L. A. F. Park et al. (Eds.): AusDM 2022, CCIS 1741, pp. 205–216, 2022.
https://doi.org/10.1007/978-981-19-8746-5_15

with advertiser's demand. Many ad networks use publishers' domains masking in order to hide the actual domain name, publisher or placement identifier. For instance, MGID (www.mgid.com) is an advertising platform that identifies a page (or an ad slot) by a UID (*unique traffic source ID*) that provides no information about its content, users visiting it nor impressions. Hence, it is hard to tell whether our ad will be well placed on a given page as we do not have access to any information about its content. This problem is further amplified by the fact that MGID connects many small publishers which attract a limited number of clicks from our campaigns, e.g., few clicks per day. In other words, there are not enough clicks to directly estimate the performance of our campaign for a given UID in any statistically reasonable way as the conversion rates (CRs) are of the order of tenths of a percent. Moreover, each day new pages appear and some other pages disappear. Specifically, when we try to block some underperforming pages the networks will deliver traffic via new sources.

Agencies and media operators are responsible for their clients' marketing campaigns. They have to rely on measurement to optimize many campaigns' metrics such as ROI, CR or profit. However, manually going through multiple campaign statistics can be very ineffective. Moreover, it is not possible to ensure by the human operators that the campaigns remain profitable on all domains, as this requires combining knowledge of many different statistics.

In this paper, we propose a novel method for enriching existing performance marketing software with a domain blocking algorithm to improve campaigns' related KPIs. We define *traffic-fingerprints* for UIDs which are defined as normalized 24-dimensional vectors representing the number of clicks collected in each hour of the day over the history of interaction with each page. Traffic-fingerprints, despite not directly related to KPIs, shall represent different types of pages and their quality of traffic. We verify this conjecture by running offline and online experiment in the following way. The traffic-fingerprints are divided into groups using specially tuned k-means clustering. Then we treat pages in each cluster jointly, i.e., we turn all of them on or off during specific hours. This approach allowed us to increase the profit of our MGID-campaigns by 53%. These proof-of-concept results were obtained via A/B tests on a system that was enriched with the above blocking algorithm.

Related Work. Understanding and modeling web traffic is a topic that has attracted research for years already [6,16]. Moreover, unsupervised learning techniques like clustering have been widely used for analysing web traffic. However, such analyses have been typically done to create patterns for user behaviour. For example, in [11] users are being segmented based on their social media advertising perceptions which reveals groups of users more susceptible to such ads. Another context for clustering users based on behaviour is web page recommendation [5] or product recommendation [18]. Although these papers use contextual data, some approaches create user profiles solely based on click-stream data [3,8,12–15]. These last approaches differ from ours in a two-fold way. First, we cluster pages and not users. Second, we use traffic-fingerprints, whereas these papers study the whole user click-streams, which in our context are not available. We note, however, that some of these papers studied behaviour patterns with respect to time [8,18] and have revealed that some user profiles are active over a specified period of days, e.g., reading news can be more often done in the morning. From a very different perspective, time patterns might be very important in predicting the performance of restaurants or other physical venues [4].

Another, use case for studying click streams is increasing the general performance of ad networks. On one hand, there are built models in order to detect and measure click-spam [2], or filter low-quality clicks that do not lead to conversions [10]. On the other hand, one might want to address the problem of fraudulent pages, i.e., content farms [9]. However, [18] revealed that advertising on content farms might be still profitable.

2 Algorithm

The main goal of the algorithm is to select web pages (domains) that are non-profitable and ads should be blocked from appearing on those web pages. The blocking might be restricted only to specific hours of the day. Due to the fact that conversions are very rare events and many domains have few clicks, we cannot wait for collecting sufficiently many clicks on every domain to estimate CR. Hence, we would like to analyse jointly some groups of domains and based on this build blocking rules (i.e., which clusters should be blocked completely or be blocked during certain hours). In order to achieve this goal we craft a new algorithm bases on combination of several unsupervised learning methods. The main novelty of this method demonstration that cluster CR/profit statistics can be used to guide us to take right decisions even for low-traffic pages based just on the similarity of traffic-fingerprints.

The algorithm consists of three steps:

1. Clustering domains based on their traffic-fingerprints using k-means together with elbow method.
2. Creating blocking rules based on statistics of the whole clusters.
3. For each day computing updated vector representations of domains and reassigning domains to clusters.
4. Blocking the domains from the selected clusters based on the cluster rules.

In such a way we are able to boost campaigns' performance by cutting off the low-quality non-profitable traffic that generates a lot of clicks throughout many different small pages but few conversions. In Sect. 3 we compare this algorithm against two baseline solutions:

noclustering - blocks the whole traffic over specific hours of day, i.e., no clustering is being used,

notime - in this approach we use our clustering algorithms but afterwords we either turn on or off the clusters completely.

We demonstrate that our algorithm is superior over these two more straightforward solutions.

Algorithm 1. Elbow method applied to k-means algorithm.

1: **function** $elbow\_kmeans(Q = [0.92, 0.95, 0.96], min = 2, max = 5)$
2:                 ▷ $Q = [q_3, ..., q_5]$ a required improvement after adding new cluster
3:     $kmeans\_clusters = compute\_kmeans(k = min)$
4:     $old\_score = compute\_kmeans\_score(kmeans\_clusters)$
5:     **for** $j = min + 1 \ldots max$ **do**
6:         $kmeans\_clusters = compute\_kmeans(k = j)$
7:         **if** $new\_score/old\_score \leq Q[j]$ **then**
8:             $old\_score = new\_score$
9:         **else**
10:             **return** $j - 1$
11:         **return** max

2.1　Step 1 – Clustering of Domains

For each domain we count clicks c_i for each hour of the day $i = 0, \ldots, 23$ over the whole history of our interaction with these domains in all our marketing campaigns. Then we normalize the values by the total number of clicks on the domain – the resulting vectors are called *traffic-fingerprints*. We define f_i for $i = 0, \ldots, 23$ as:

$$f_i = \frac{c_i}{\sum_{j=0}^{23} c_j}.$$

The next step is described in Algorithm 1. We cluster 24-dimensional traffic-fingerprints using the k-means algorithm. For determining the optimal number of clusters, for each ad network we start with 2 clusters and then gradually increase the number as far as adding another cluster does not give a relevant improvement to the score, according to the so-called Elbow method [7,17].

2.2　Step 2 – Creating Blocking Rules

For each cluster we study the dependence of profit on the hour. The shape of these graphs looks similar for different networks, but for some, both clusters are above profitability around the clock, and for others, the weaker clusters "dive" below profitability for some part of the day. Using this dependence we generate blocking rules that contain hours with negative profit. The procedure for creating blocking rules is given as Algorithm 2.

It was always the case that the night hours are weaker, therefore it is possible to set two points – in the evening and in the morning – which are the limit of profitability. This is important as turning the campaign off and on for big clusters of domains is a time-consuming process and ad networks need some period of time to re-optimize campaigns due to changes in the campaign's configuration. Therefore, it only makes sense that it is done sparingly (e.g., once per day).

Algorithm 2. Creates blocking rules for all domains in a cluster.

1: **function** $create\_blocking\_rules(d, X, P)$
2:                         ▷ d – number of domains
3:                         ▷ $X = [f_1, ..., f_d]$ – traffic-fingerprints
4:                         ▷ $P = [p_1, ..., p_d]$ – p_i is a 24-length hourly profit vector for domain i
5:        $k = elbow\_kmeans(Q, min = 2, max = 5)$
6:        $kmeans\_clusters = compute\_kmeans(X, k = k)$
7:        $blocking\_rules = \{\}$
8:        **for** i=1...k **do**
9:                $cluster\_members = get\_cluster\_members(kmeans\_clusters)$
10:                                ▷ the set of domains belonging to the cluster i
11:                $mean\_profit = average([p_j : j \in cluster\_members])$
12:                        ▷ average profit from all domains from the cluster (24-length vector)
13:                $smoothed\_mean\_profit = moving\_average(mean\_profit, 4)$
14:                        ▷ to reduce variance of mean_profit compute 4 hour moving averages
15:                $profitable\_hours = (smoothed\_mean\_profit > 0)$
16:                **if** $all(profitable\_hours == True)$ **then**
17:                        $rule = 1$
18:                **else**
19:                        **if** $all(profitable_hours == False)$ **then**
20:                                $rule = -1$
21:                        **else**
22:                                $rule = [idxmax(profitable\_hours) + 1, idxmin(profitable\_hours)]$
23:                $blocking\_rules[i] = rule$

2.3   Step 3 – Reassigning Domains to Clusters

Having computed clusters of domains on historical data, we fix clusters' centers (in the k-means algorithm each cluster is defined by its centroid). Then, while we are collecting new traffic on domains, we update traffic-fingerprints for domains (in the same way as in Step 1). Finally, for each domain we select the closest cluster (in the Euclidean metric). We decided to not redo clustering after new data is collected. Our offline simulations suggest that this does typically change the final blocking decisions significantly. However, it does require monitoring the number and the character of clusters found and matching them to the original ones which is an unnecessary complication.

What is important, when a domain is blocked during some hours on a given day, we do not include statistics for this day in the updated traffic-fingerprints. Otherwise, we would disturb the representation of this domain's natural traffic and resulting fingerprints would not represent true traffic that would be normally attracted by the domain.

3   Offline Experiments

Experimental Environment. We performed experiments with our algorithm on several ad networks. For each ad network we split the data into train and test sets.

Table 1. Basic statistics for obtained clusters in examined ad networks. Increase of profit shown here was obtained in offline test.

Ad network	Silhouette score	Increase of profit	Cluster	Blocking rule
MGID	0.14	41%	0	Blocked 16 - 24
			1	Not blocked
			2	Blocked
			3	Blocked
Exoclick	0.18	211.3%	0	Not blocked
			1	Blocked 20 - 5
Content Stream	0.19	4.9%	0	Blocked
			1	Non blocked
Taboola	0.14	7.6%	0	Not blocked
			1	Blocked 15 - 5
Traffic Stars	0.12	56.7%	0	Blocked 22 - 6
			1	Blocked

The information about data sets and the split is given in Table 3 and Table 4. For clustering, we selected only domains with at least 50 clicks to ensure the robustness of domains' vectors. On the train set we selected the optimum number of clusters and made decisions which clusters at which hours should be blocked (steps (1) and (2) of the algorithm). Then, we proceeded according to the chosen strategy on the test set - by recomputing traffic-fingerprints, reassigning domains to clusters (step (3)) and simulating clusters' blocking. Finally, we compared the statistics obtained in the simulation with the real statistics.

We should notice that in this approach the results can be inaccurate, because in reality blocking domains has an impact on the behaviour of ad networks, i.e., after blocking some domains we will observe additional traffic on other domains (because the ad network could aim to use the whole available budget). Hence, only online experiments (described in the next section) can give us definitive results. However, in the offline experiments we were able to test a wider range of traffic and check if the clusters' patterns identified in one ad network are similar to clusters in other ad networks (and that we avoided overfitting to a specific ad network's strategy).

Results. Table 1 shows the basic statistics obtained in offline experiments for five examined ad networks. We can see the selected number of clusters with the silhouette score and the rules for blocking that were obtained on the training set. The column *Increase of profit* shows us the gain after applying the rules on the test set. For every ad network the increase was positive, however, as expected the results vary for different networks. We note that for MGID network the strategy is rather nontrivial, as it includes four clusters with different behaviour: one is blocked only overnight, one is not blocked a all, and two are blocked completely.

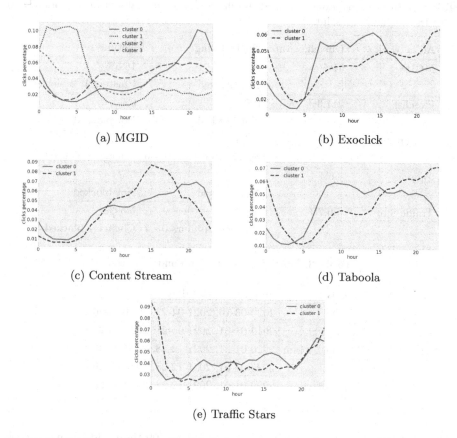

Fig. 1. Traffic-fingerprints for identified clusters' centers for studied ad networks. A clear difference of daily traffic patterns is seen between the clusters.

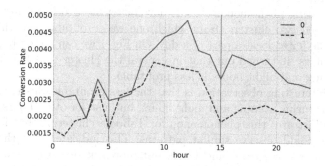

Fig. 2. Conversion rate, in dependence on hour of day, for two clusters of Taboola clustering. Vertical lines show hours of (un)blocking cluster 1.

Table 2. Increase of profit obtained in offline test for our algorithm in comparison with the two baseline solutions.

Ad network	Our algorithm	Noclustering	Notime
MGID	41%	0%	16.5%
		Not blocked	Clusters 1, 2 blocked
Exoclick	211.3%	0%	0%
		Not blocked	Not blocked
Content Stream	4.9%	−2.4%	4.9%
		Blocked in hours 1 - 6	Cluster 0 blocked
Taboola	7.6%	0%	0%
		Not blocked	Not blocked
Traffic Stars	56.7%	46.2%	5.59%
		Blocked in hours 0 - 7	Cluster 1 blocked

Table 3. Split of data into train and test sets.

Ad network	Start, split, end date
MGID	2020-03-16, 2021-04-23, 2021-06-30
Exoclick	2019-01-01, 2021-05-21, 2021-06-20
Content Stream	2020-01-01, 2021-12-28, 2022-01-26
Taboola	2020-01-01, 2021-12-28, 2022-01-26
Traffic Stars	2020-01-01, 2021-12-28, 2022-01-26

The clustering for the test set contains more than 300 thousands domains each with just 18 clicks on average. It is hard to imagine that a human operator would be able to generate and maintain such set of rules. In Table 2 we compare our algorithm against the two baseline solutions, i.e., noclustering and notime. We note that these ideas separately lead to increase of profit only in some cases.

An interesting phenomenon is that our clustering algorithm, in most of the cases generated two clusters, from which one was profitable and the second – non-profitable for at least a part of a day. On Fig. 1 we can see daily patterns of clusters' centers. It turned out that in general the cluster that has more traffic during evening and night hours is less profitable.

A very interesting observation is that the less profitable clusters have lower CR not only during evening and night hours, but also for other hours of the day. The Fig. 2 illustrates this phenomenon for Taboola clustering, but for other ad networks we could observe similar patterns. This indicates that the identified clusters are not just some coincidence, but represent domains with a different type of daily activities that can be identified by their traffic and thus have a different value for e-commerce.

Table 4. Information about train and test sets used in offline experiments.

Ad network	Budget (train/test)	Clicks (train/test)	Domains (train/test)
MGID	54,767/18,997	5,416,140/825,672	304,996/62,036
Exoclick	1,332,013/35,436	41,483,870/1,524,199	22,052/5,938
Content Stream	218,870/16,635	4,029,076/427,148	2,246/732
Taboola	295,327/13,810	4,542,487/223,299	8,540/1,434
Traffic Stars	195,445/1,621	6,331,319/58,005	1,630/807

We have suspected that the revealed clusters might correspond to some topical groups of pages. Thus we have investigated contextual data delivered the networks which make this data available, i.e., all but MGID. We were unable to see any topical nor geographical characteristics of the clusters. In our case, these networks did provide us with very homogeneous traffic (e.g., Content Stream delivered only news-like local web pages, whereas other networks delivered various global domains). Hence, we were not able to spot any significant topical nor geographical difference between the two clusters used by our algorithm.

4 Online Experiments

We executed online experiments with our algorithm on several campaigns on the MGID platform. These experiment were executed in a hybrid system, where for strategic decisions were taken by human operators, whereas our algorithm was responsible for blocking rules. Thus the campaign parameters were configured by human operators who selected such values as a market, advertisements, appropriate stakes per ad and so on. Then during the whole examined period the operators have been monitoring the campaigns' profitability and have been performing necessary actions such as modifying stakes or changing ads. The operators based their decisions on their experience and simple rules like "if the ad is not profitable, remove it and add a new one with a higher stake".

The exact A/B tests were impossible, because the traffic for our campaigns is provided by the ad network and we are not able to directly choose the web pages, where we will display ads. So we are not able to create two exactly the same campaigns.

Hence, we decided to select a subset of all active campaigns and to run the algorithm on this subset (a similar approach for an experimental design was proposed, e.g., in [1]), while the other part of campaigns was managed as before by human operators. This way, we could compare the periods before and after running optimization, and also the results for two campaign groups.

When we introduced the algorithm for blocking clusters of domains, the operators have been acting on the all the campaigns as before. Hence, during the A/B tests part of the campaigns were optimized semi-automatically, i.e., jointly by human operators and our algorithm, whereas the other part was managed by humans only. We note that the absolute value the result of the campaigns depended highly on the correctness of the human actions. Depending on these factors we might observe fluctuations in the results, but what maters for us is not he absolute value before, during or after the test, but the difference between the two test groups.

We have been monitoring all campaigns' performance on MGID from the beginning of June 2021 (the number of active campaigns varied before 20 and 40, since the campaigns were constantly paused and restarted by the operators due to market changes). Then, on the 7th of July we randomly selected about half of the campaigns and began the optimization by the algorithm. Finally, on the 9th of September we turned on the algorithm on all active campaigns.

Figure 3 shows 14-days moving averages of the total profit per click for optimized and non-optimized campaigns. Vertical lines indicate the days when the algorithm was turned on for some campaign groups. Despite high variability of the metric, we can observe a noticeable jump in the performance after turning on the optimization in both time stamps. You should note that the red optimized line is systematically above the blue non-optimized one during almost the whole A/B tests period (i.e., not including the starting period where the whole system adopts to the new strategy). This clearly indicates that increased performance of the optimized system.

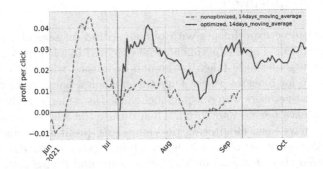

Fig. 3. Average profit per click for all campaigns in A/B tests. The dashed line shows the performance of the system without automatic cluster blocking, whereas the solid line shows results with it. Vertical lines show the days when the algorithm was turned on for some campaigns: 7th of July half of active campaigns and 9th of September all campaigns.

In Table 5 we can see the other metrics for 37 campaigns that were optimized for 14 days only by operators and 14 days later also by the algorithm. After turning on the algorithm, total profit increased by about 53.4%, while the profit per click increased by 37%. What is important, despite temporary blocking some web pages, the total volume of the traffic has not decreased (actually, the number of clicks even slightly increased). In other words, the MGID network ended up delivering more and higher quality traffic. This happened thanks to step 3 of the algorithm, which is able to handle new pages even with very low traffic, i.e.,

Table 5. Statistics for 14 day periods before and after blocking was introduced. The profit increased by more than 50%.

Blocked period	Clicks	Conversions	CR %	Profit	ROAS	Profit per click
False	186,528	760	0.41	3931.55	1.23	0.021
True	208,773	905	0.43	6030.20	1.31	0.029

just several clicks. The A/B tests were executed over the time period of slightly more than two months, thus giving the MGID management algorithms more than enough time to adopt and optimize its adversarial strategy.

5   Conclusions

We demonstrated that in order to effectively manage blocking domains in ad networks, it is enough to collect just 50 clicks per single domain. The proposed domains' blocking algorithm allowed us to increase profit by over 40% both in offline simulations on historical data and as well as online on real e-commerce campaigns. We suspect that such improvement was not caused by traffic management algorithms in the MGID ad network, but is a result of real differences in traffic quality between domains. This is testified by similar shapes of clusters' centroids traffic-fingerprints in all considered ad networks.

In the future work we plan to extend our approach as follows: training supervised learning models, that using domains' traffic-fingerprints predict domain profitability. Next, use these models in order to block domains predicted as unprofitable. However, this approach will have many challenges, like interpretability and building blocking rules that do not turn on and turn off domains too often. In this aspects the proposed clustering method is more understandable from a business point of view and easier to use.

Furthermore, our current solution automates only partially the tasks necessary to run profitable e-commerce online campaigns. The next steps should focus on the full substitute of human campaign operators by sets of algorithms. Still, the presented solutions would form one of the most innovative aspects of the system we develop.

References

1. Agarwal, D., Ghosh, S., Wei, K., You, S.: Budget pacing for targeted online advertisements at linkedin. In: Proceedings of the 20th ACM SIGKDD International Conference on Knowledge Discovery and Data Mining, pp. 1613–1619 (2014)
2. Dave, V., Guha, S., Zhang, Y.: Measuring and fingerprinting click-spam in ad networks. ACM SIGCOMM Comput. Commun. Rev. **42** (2012). https://doi.org/10.1145/2342356.2342394
3. Dixit, V.S., Gupta, S.: Personalized recommender agent for E-commerce products based on data mining techniques. In: Thampi, S.M., et al. (eds.) Intelligent Systems, Technologies and Applications. AISC, vol. 910, pp. 77–90. Springer, Singapore (2020). https://doi.org/10.1007/978-981-13-6095-4_6

4. D'Silva, K., Noulas, A., Musolesi, M., Mascolo, C., Sklar, M.: Predicting the temporal activity patterns of new venues. EPJ Data Sci. **7**(1), 1–17 (2018). https://doi.org/10.1140/epjds/s13688-018-0142-z

5. Fabra, J., Álvarez, P., Ezpeleta, J.: Log-based session profiling and online behavioral prediction in e-commerce websites. IEEE Access **8**, 171834–171850 (2020). https://doi.org/10.1109/ACCESS.2020.3024649

6. Ihm, S., Pai, V.S.: Towards understanding modern web traffic. In: Proceedings of the 2011 ACM SIGCOMM Conference on Internet Measurement Conference, pp. 295–312. IMC 2011, Association for Computing Machinery, New York, NY, USA (2011). https://doi.org/10.1145/2068816.2068845

7. Ketchen, D.J., Shook, C.L.: The application of cluster analysis in strategic management research: an analysis and critique. Strateg. Manag. J. **17**(6), 441–458 (1996). https://doi.org/10.1002/(SICI)1097-0266(199606)17:6<441::AID-SMJ819>3.0.CO;2-G

8. Kleppe, M., Otte, M.: Analysing and understanding news consumption patterns by tracking online user behaviour with a multimodal research design. Digital Sch. Humanit. **32**(2), ii158–ii170 (2017). https://doi.org/10.1093/llc/fqx030

9. Luh, C.J., Wu, A.: Is it worth to deliver display ads on content farm websites. J. Comput. **30**, 279–289 (2019)

10. Mungamuru, B., Garcia-Molina, H.: Managing the quality of CPC traffic. In: Proceedings of the 10th ACM Conference on Electronic Commerce, pp. 215–224. EC 2009, Association for Computing Machinery, New York, NY, USA (2009). https://doi.org/10.1145/1566374.1566406

11. Nasir, V.A., Keserel, A.C., Surgit, O.E., Nalbant, M.: Segmenting consumers based on social media advertising perceptions: how does purchase intention differ across segments? Telematics and Inform. **64**, 101687 (2021). https://doi.org/10.1016/j.tele.2021.101687, https://www.sciencedirect.com/science/article/pii/S073658532100126X

12. Pai, D., Sharang, A., Yadagiri, M.M., Agrawal, S.: Modelling visit similarity using click-stream data: a supervised approach. In: Benatallah, B., Bestavros, A., Manolopoulos, Y., Vakali, A., Zhang, Y. (eds.) WISE 2014. LNCS, vol. 8786, pp. 135–145. Springer, Cham (2014). https://doi.org/10.1007/978-3-319-11749-2_11

13. Singh, H., Kaur, P.: An effective clustering-based web page recommendation framework for e-commerce websites. SN Comput. Sci. **2**, 339 (2021). https://doi.org/10.1007/s42979-021-00736-z

14. Su, Q., Chen, L.: A method for discovering clusters of e-commerce interest patterns using click-stream data. Electron. Commer. Res. Appl. **14**(1), 1–13 (2015). https://doi.org/10.1016/j.elerap.2014.10.002, https://www.sciencedirect.com/science/article/pii/S1567422314000726

15. Thiyagarajan, R., Kuttiyannan, D.T., Ramalingam, R.: Recommendation of web pages using weighted k-means clustering. Int. J. Comput. Appl. **86** (2013). https://doi.org/10.5120/15057-3517

16. Thompson, K., Miller, G.J., Wilder, R.: Wide-area internet traffic patterns and characteristics. IEEE Netw. **11**, 10–23 (1997)

17. Thorndike, R.L.: Who belongs in the family? Psychometrika **18**(4), 267–276 (1953). https://doi.org/10.1007/BF02289263

18. Vanessa, N., Japutra, A.: Contextual marketing based on customer buying pattern in grocery e-commerce: the case of bigbasket.com (India). Asean Market. J. 56–67 (2018)

Attractiveness Analysis for Health Claims on Food Packages

Xiao Li[1], Huizhi Liang[2(✉)], Chris Ryder[3], Rodney Jones[3], and Zehao Liu[3]

[1] University of Aberdeen, Aberdeen, UK
x.li.12@aberdeen.ac.uk
[2] Newcastle University, Newcastle upon Tyne, UK
huizhi.liang@newcastle.ac.uk
[3] University of Reading, Reading, UK
{c.s.ryder,r.h.jones,zehao.liu}@reading.ac.uk

Abstract. Health Claims (Health Claims) on food packages are statements used to describe the relationship between the nutritional content and the health benefits of food products. They are popularly used by food manufacturers to attract consumers and promote their products. How to design and develop NLP tools to better support the food industry to predict the attractiveness of health claims has not yet been investigated. To bridge this gap, we propose a novel NLP task: attractiveness analysis. We collected two datasets: 1) a health claim dataset that contains both EU approved Health Claims and publicly available Health claims from food products sold in supermarkets in EU countries; 2) a consumer preference dataset that contains a large set of health claim pairs with preference labels. Using these data, we propose a novel model focusing on the syntactic and pragmatic features of health claims for consumer preference prediction. The experimental results show the proposed model achieves high prediction accuracy. Beyond the prediction model, as case studies, we proposed and validated three important attractiveness factors: specialised terminology, sentiment, and metaphor. The results suggest that the proposed model can be effectively used for attractiveness analysis. This research contributes to developing an AI-powered decision making support tool for food manufacturers in designing attractive health claims for consumers.

Keywords: Attractive analysis · Health claims · Learning-to-rank · Consumer preference prediction

1 Introduction

Health claims on food packages (e.g. Fig. 1) are statements used to describe the relationship between the nutritional content and the health benefits of food products for product promotion. Food manufacturers are increasingly including health claims on their packages [10]. Recent research shows that the presence of such claims on packages generally has a positive impact on consumers' perceptions of the healthiness of products and their willingness to buy them [1]. To protect consumers from being deceived or misled, health claims are strictly regulated in most places in the world. In the European Union, the use of health claims on food packages and in other marketing materials is

L. A. F. Park et al. (Eds.): AusDM 2022, CCIS 1741, pp. 217–232, 2022.
https://doi.org/10.1007/978-981-19-8746-5_16

Table 1. Example Health Claims approved by Regulation (EC) 432/2012 and example Health Claims collected from food packages

(a) Approved Health Claims	(b) Revised Health Claims for real commercial use
Vitamin A contributes to the normal function of the immune system	High in vitamin A which supports the normal function of the immune system
Vitamin C contributes to the protection of cells from oxidative stress	Vitamin C contributes to antioxidant activity in the cells (to help protect them from damaging oxidative stress)
Potassium contributes to the maintenance of normal blood pressure	Potassium plays role in maintaining normal blood pressure
Selenium contributes to the normal function of the immune system	Selenium helps maintain your immunity system

governed by Regulation (EC) No. 1924/2006. According to the regulation, manufacturers that sell their products within the European Union may only include health claims that have been approved for use by the European Commission based on the verification of their scientific substantiation by the European Food Safety Authority (EFSA). However, health claims approved by EFSA are written in dense scientific language which is sometimes difficult for consumers to understand. Rewording of approved health claims is allowed as long as the revised claim has the same meaning as the approved claim, as stated in Regulation (EC) 432/2012. In practice, food manufacturers often attempt to rewrite approved health claims in order to communicate the health benefits of their products in an easy-to-understand, unique, and attractive way. Table 1 shows examples of approved and revised health Cclaims. Approved health claims (see Table 1(a)) consist of information about 1) the nutrient contained in the product, and 2) the health benefits of the nutrient, following a relatively standard template, for example:

[nutrient] *contributes to* **[health benefit]**

Table 1(b) shows examples of revised health claims. We can see that manufacturers usually revise health claims by focusing on certain linguistic features, for example, word choice, syntactic structure, emotional valiance, etc. Traditionally, nutritionists, marketers, and lawyers hired by food companies work together to formulate and vet

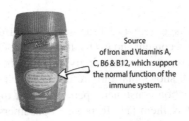

Source of Iron and Vitamins A, C, B6 & B12, which support the normal function of the immune system.

Fig. 1. An example health claim on a food package.

these revised versions of health claims . Then they often conduct user studies or A/B testing for each revised version. Currently, there is no systematic research attempting to link the attractiveness of health claims to specific linguistic characteristics.

In this paper we propose to analyze the attractiveness of Health Claims on food packages by developing a computational prediction indicator to determine how attractive a health claim is likely to be for consumers. Specifically, we compute a Consumer Preference score (CP score) for specific health claims. The higher the score, the more likely the health claim will be welcomed by consumers.

The challenges of this study include: (1) The difficulty of defining linguistic criteria to measure the attractiveness of health claims. Thus, one cannot evaluate health claims based on pre-defined guidelines. (2) The range of linguistic variables involved. Since EU regulations stipulate that the literal meaning of revised health claims should be similar or the identical to corresponding approved health claims, semantic meaning is unlikely to be the main factor in determining whether or not a revised health claim is more or less attractive than its corresponding approved claim. Other factors including syntax and pragmatics are likely to be more important. (3) The difficulty of obtaining negative labels (i.e., very unattractive health claims), since the revised health claims on food packages have been designed by experts with the explicit aim of attracting consumers. (4) The difficulty of obtaining human evaluations of health claims that substantially diverge from the approved claims, since presenting unapproved or inaccurate health claims to people might be considered unethical.

To address the above challenges, we first collected two datasets, which we discuss in Sect. 3.1. Then we designed a consumer preference prediction model, as discussed in Sect. 3.2. The evaluation of the proposed model is given in Sect. 4. Finally, in Sect. 5, we analyse the attractiveness factors of health claims through three case studies based on our model. The contribution of this work can be summarised as follows: (1) We have framed a new application area or task for NLP techniques: attractiveness analysis for health claims in the food industry; (2) We have collected the first health claim datasets for attractiveness analysis; (3) We propose a novel model for studying the attractiveness of health claims, which has achieved high accuracy in our evaluation; (4) We have demonstrated effective means of investigating attractiveness factors in health claims using the model.

2   Related Work

Although health claims are widely used in the food industry, there is surprisingly little scientific research available to support food companies in making decisions about how to formulate Health Claims. Previous research, [17] finds that consumers prefer food products with health claims on the packaging compared to products without health claims, and give greater weight to the information mentioned in health claims than to the information available in the Nutrition Facts panel. [18] suggests that consumers have individual differences in their preferences for health claims, but these differences can also be attributed to consumer cultural differences. Regarding the content of health claims, [9] try to improve health Claims by adopting the Decisions Framing [19]; a study by [5] states that foods that emphasise healthy positive contributions to life (referred to as life marketing) are more attractive to consumers than foods that emphasise avoidance of disease (death marketing); [20] finds that consumers may be reluctant to try products whose health claims include unfamiliar concepts because consumers tend to evaluate them as less credible. However, all of this research focuses on the relationship among food products, health claims, and consumer attitudes rather than focusing on the the specific linguistic features of health claims.

Currently, NLP techniques have been widely applied in linguistic studies [8]. NLP allows us to use quantitative research methods to study abstract linguistic phenomena. NLP is particularly popular in studeis of syntax and pragmatics [3], e.g., dependency parsing [13], metaphor processing [15], and sentiment analysis [6]. However, to the best of our knowledge, applying these NLP tools to the analysis of the attractiveness of health claims is still new.

3   Consumer Preference Prediction of Health Claims

Learning consumer preferences for health claims is defined as learning a function ($f(\cdot)$) that maps health claims to a real number (called the Consumer Preference score or CP score for short, denoted by u). The real number indicates the degree of consumer preference for an input health claim text (denoted by x). The larger the value of u, the more attractive the health claim is assumed to be.

$$f(x) = u \tag{1}$$

3.1   Dataset Collection

Our dataset collection had two stages, which generates a health claims dataset with pair-wise customers preference labels. In the first stage, we collected a large number of real-life health claims from the food products sold in EU supermarkets. Since vitamins and minerals are everyday nutrients that are familiar to consumers and can be found in many food products, we only used the Health Claims for vitamins and minerals in our research to better control the experimental variables. At this stage, we collected a total of 4200 Health Claims text.

In the second stage, we used scenario-based experiments to observe consumers' (virtual) purchase intentions, based on the collected Health Claims. Figure 2 shows an example task in our experiments. In the experiments, a subject is asked to help Alex to choose a food product as a gift for Sam. We used neutral names (Alex and Sam) and pronouns "them" rather than "he" or "she" to avoid gender bias. The decision was made according to two random-paired health claims. Each subject was asked to complete 20 tasks. The options in each task are displayed as a gift pair consisting of the random health claims that we collected. Because of the randomness, the questionnaires of one subject might be different from one another. We recruited 200 subjects from the EU and the UK[1] via Amazon MTurk. 183 of them were valid, i.e. those who completed the questionnaire within 30 min and spent more than 3 s on each task on average. Totally, we gathered $183 \times 20 = 3660$ answers for the paired health claims with a selected preferred label for each pair.

---

[1] The locations of subjects were filtered by MTurk.

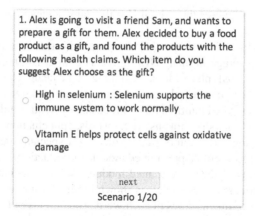

Fig. 2. A scenario task example for the Consumer Health Claim Preference data collection.

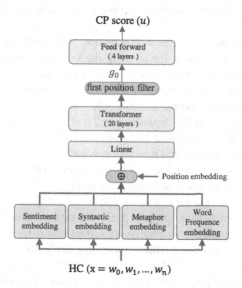

Fig. 3. The architecture of our proposed model (i.e. $f(\cdot)$ in Eq. 1).

3.2 Prediction Model

Our model is used to learn $f(\cdot)$ in Eq. 1. The input of the model is a health claim, and the output is the Consumer Preference score of the health claim. The model aims to focus as much as possible on the preferences for linguistic features in health claims rather than their literal meanings. First, as the literal meanings of revised health claims are governed by EU regulations, revised health claims should mainly have the same semantic meaning as the corresponding approved health claims. Second, previous research [5] suggests that consumer preferences for health claim are affected by deep-level linguistic feature, such as sentiment factors [5] or unfamiliar concepts [20]. If a model focuses on specific words but ignores the overall Health Claim sentence, it may lack practical significance.

The model adopts a Transformer [21] based architecture (see Fig. 3), since it has demonstrated its advantage in many NLP tasks. The input is a health claim sentence with a special token ([sos]) placed at the beginning of the health claim sentence. Unlike common NLP practices that mainly embed semantic information of the input text in vector space, we considered multiple factors focusing on syntax and pragmatic factors.

Specifically, we encoded tokens (i.e. words, punctuation etc.) denoted as $(w_0, ..., w_n)$ of the input as vectors $(h_0, ..., h_n)$ via a specially designed embedding layer which consists of multiple pre-trained models in syntactic and pragmatic analysis tasks [7]. The outputs of these pre-trained models were concatenated together as the input of the transformers of our model, as shown in Fig. 3. First, we used the model of [7] to calculate the sentiment scores of each word. The outputs of this sentiment model are 4-dimensional vectors including the overall sentiment score (in $[-1, 1]$), the positive, neutral, and negative scores (in $[0, 1]$). The model of [14] was employed to generate metaphoricity scores that indicate the metaphoric possibility for each word (i.e., a real number in $[0, 1]$).

Second, we parsed health claims with a python NLP module (spaCy) to obtain the dependency relationship features for forming the syntactic embedding in Fig. 3. For each word, we adopted the distance between a word and its dependent word as the partial syntactic information. Here, we encode the distances by using the same method of the learnt positional embedding. For example, suppose w_1 depends on w_2 and w_2 depends on w_6 which means the distances are $2 - 1 = 1$ and $6 - 2 = 4$, we encode 1 and 4 by the way of the positional embedding as the partial embedding for w_1 and w_2. The edge tags (i.e. grammatical relations annotating the dependencies e.g. *dobj* for direct objects, *conj* for conjunct, etc.) were also employed to represent the syntactic relationship, which was encoded with one-hot encoding.

Finally, we used the logarithm of word frequency and the tf-idf values as word frequency embedding features to assign more weight to rare words and concepts. The word frequencies and the document frequency were obtained from Google Books Ngram dataset[2]. We also used the learnt position embedding to identify the word positions. The final embedding vector (h_i for word w_i) is the concatenation of all the 5 kinds of embedding features. The embedding vector for [sos] token was set to a zero vector. After the embedding layer, $h_0, ..., h_n$ were fed into a linear layer to compress their feature maps to 32-dimension, before feeding into a 20-layer transformer encoder. According to [2], narrowing the width (32-dimension) of a model and increasing its depth (20-layer) can mitigate overfittings. Then, the output vector for the special token [sos] was used as the pooling hidden vector (g_0), feeding to a 4-layer feed forward neural network, obtaining the final Consumer Preference score (u).

The training process adopted a Learning-to-Rank strategy [4] for learning global sorting scores (i.e. the Consumer Preference score u) of health claims from paired examples. Given two paired health claims (x_r^+, x_r^-), where x_r^+ is preferred to x_r^- in an health claim pair (r). The training process is to force the Consumer Preference score of x_r^+ to be higher than that of x_r^-, that is to let $u_r^+ > u_r^-$ (Eq. 2).

---

[2] https://storage.googleapis.com/books/ngrams/books/datasetsv3.html.

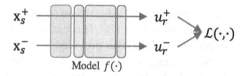

Fig. 4. The training process (Eq. 3). Two paired health claims are fed to the model iteratively, where the model learns to yield a higher score (u_r^+) for the preferred health claim than the score of the other health claim (u_r^-) of a given health claim pair.

$$u_r^+ = f(x_r^+)$$
$$u_r^- = f(x_r^-)$$

(2)

Unlike the usual practices of Learning-to-Rank, our model adopted an exponential loss function (Eq. 3) R denotes the training dataset and $|R|$ is the size of R. It forces the value of $u_r^- - u_r^+$ to be as small as possible just as the usual Learning-to-Rank approaches. However, when $u_r^- - u_r^+ > 0$ (i.e., wrong predicted preference order of x_s^+ and x_s^-), our exponential loss yields extra penalties.

$$\mathcal{L} = \frac{1}{|R|} \sum_{r \in R} e^{u_r^- - u_r^+}$$

(3)

In the training process, x_s^+ and x_s^- in a paired health claims are fed into the model iteratively, so that when the model predicts Consumer Preference score for one health claim, another health claim is not considered (Fig. 4). According to the loss (Eq. 3), the gradients on the model depend on the relative Consumer Preference scores (i.e. $u_r^- - u_r^+$) rather than the individual values of u_r^- and u_r^+. If and only if the model predicts the wrong order for two health claims (x_r^+ and x_r^-), the model parameters will be updated significantly.

4   Evaluation and Results

We examined our model with 50-cross validation. Random 80% of data was used for training, and 20% was used for testing. The model was trained for 500 epochs with a batch size of 256. The accuracy of automatic evaluation was measured by the averaged accuracy of the last 10 epochs. We used an AdamW optimiser with the learning rate of 10^{-5}, $\beta_1 = 0.9$, $\beta_2 = 0.999$, and $\epsilon = 10^{-7}$, where the weight decay was 10^{-2}. The example health claims with the predicted Consumer Preference scores are ranked in Table 3. We introduced a baseline model with a learnt word embedding layer with randomly initialised weights layer (named Learnt embedding in Table 2), and a BERT baseline (named Bert). We allowed weight updating for Bert and the Learnt embedding layer during training (Bert with fine tune). Since we proposed the exponential-based loss function, we also compared it with the original Learn-to-Rank loss which is the Cross-Entropy Loss (i.e. the model adopting our embedding method and Cross-Entropy Loss). We tried to test the utilities of our embedding layer.

As seen in Table 2, our proposed model achieved an accuracy of 76% on the testing set, outperforming the baselines by at least 8%. This can be explained by the fact that the pre-trained models in our embedding layer provided extra syntactic and pragmatic knowledge. Reflecting on the accuracy scores, the BERT baseline yielded a high score on the training set but low on the testing set. We infer that the BERT model exposes overfitting on the training set. This can be explained by the fact that BERT provides rich semantic features, but they may not be suitable for the attractiveness analyses of Health Claims. In addition, the Learnt embedding baseline seems not to have learnt enough appropriate knowledge for the prediction. This is probably due to the fact that the size of our training data set was not large enough for this kind of model.

Table 2. The automatic evaluation results for the consumer preference prediction. The accuracy is the mean of the accuracy scores of 50-cross-validation.

	Training set accuracy	Testing set accuracy
Learnt embedding	0.86	0.66
Bert	0.95	0.68
CrossEntropy loss	0.83	0.73
Proposed model	0.84	**0.76**

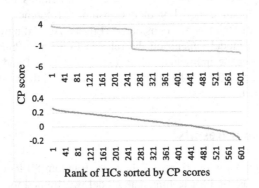

Fig. 5. The changing of the Consumer Preference scores on the sorted Health Claims. Yellow line denotes that of Cross Entropy Loss; Blue line denotes that of our exponential loss. (Color figure online)

We also compared the influence of our proposed loss function (Eq. 3) with the original Learn-to-Rank loss (i.e. the Cross-Entropy Loss). By respectively using the proposed model and the baseline model using Cross-Entropy loss (in Table 2), we scored all the Health Claims and sorted them according to the scores. After that, we investigated how the scores changed (Fig. 5) when the health claim rank changed. We can see that when the model is trained with our exponential loss, the Consumer Preference

Table 3. The top 5 Health Claims and the bottom 5 Health Claims, sorted by predicted Consumer Preference scores. Empirically, we could see that the top 5 Health Claims are much more attractive than the bottom 5 Health Claims.

Rank	CP score	Health Claims
1	0.266	It is a source of phosphorus: Phosphorus helps ensure the normal energy metabolism
2	0.261	Vitamin C helps to support a healthy immune system
3	0.260	High in vitamin B12: Vitamin B12 supports the immune system to function normally
4	0.259	High in vitamin A: Vitamin A supports the immune system to function normally
5	0.257	Naturally high in vitamin C: Vitamin C helps to protect cells from oxidative stress
...	...	...
597	−0.199	Calcium contributes to normal muscle function
598	−0.199	Magnesium contributes to normal muscle function
599	−0.199	Zinc contributes to normal macronutrient metabolism
600	−0.200	Potassium contributes to normal muscle function
601	−0.202	Zinc contributes to normal cognitive function

Table 4. The human evaluation results against null hypothesis $\mathbf{H_G}$. The p-values are calculated via Mann-Whitney U test.

Hypothesis	Group	Ratio of chosen	p-value
$G_H \succ G_L$	G_H	.82	$< 10^{-5}$
	G_L	.18	

scores show a smooth change. While it is trained by the Cross-Entropy loss, there is a sharp drop in the Consumer Preference scores. This drop may cause the model with Cross-Entropy loss to perform slightly worse than the proposed model.

In addition to the automatic evaluations, we also conducted human evaluations. We collected the top 50 health claims (denoted by G_H) that had the highest Consumer Preference scores and the last 50 health claims with the lowest scores (denoted by G_L). We conducted human evaluation by pairing each health claim in G_H with a randomly selected health claim in G_L. We meant to test whether the health claims with high Consumer Preference scores (G_H) were more preferred than the health claims with low Consumer Preference scores (G_L) via human evaluations. Based on the same experimental method in Sect. 3.1, 20 subjects participated in this survey, and each of them completed 8 tasks. Our null hypothesis ($\mathbf{H_G}$) was that health claims in G_H and G_L have no difference regarding consumer preferences. The results (Table 4) show the subjects significantly prefered (denoted by '$\succ$') the top-ranking health claims (G_H) compared to the low-ranking health claims (G_L); the p-value $< 10^{-5}$ rejects $\mathbf{H_G}$. Thus, a health claim with a high Consumer Preference score is more likely preferred by a consumer than a low Consumer Preference score health claim.

5 Case Studies

This section describes our investigation into the linguistic factors affecting the attractiveness of a health claim through case studies. We conducted two kinds of case study from the perspectives of local factors and global factors, which was to: (1) understand the important factors that determine whether a health claim is attractive or not; (2) verify whether the results, given by our model, are consistent with the consumer survey results. The findings will help manufacturers to automatically evaluate their health claims before marketing.

5.1 Specialised Terminology Factors

[20] suggests that unfamiliar concepts may prevent consumers from buying food products. [16] found that the use of specialised terminology can reduce the number of citations of papers in the domain of cave research. In light of this, we analysed the impact of specialised terminology (jargon) in health claims. Specialised utterances are local features for sentences; they are mainly related to the academic names of nutritional ingredients. E.g., *thiamin* is also known as *Vitamin B1*. The following example shows two health claims with different nutrient terminologies.

> ***Thiamin*** *helps support a healthy heart*
>
> *Vitamin B1 helps support a healthy heart*

We focused on B vitamins, e.g., *thiamin* vs. *Vitamin B1*, *riboflavin* vs. *Vitamin B2*, and *pantothenic acid* vs. *Vitamin B5*, because they can be easily found in food products. Table 5 shows the statistics of each item in our collected health claim dataset.

Table 5. The statistics of the specialised names and common names for B vitamins in our collected dataset.

Specialised utterance		Common utterance	
Utterance	count	Utterance	count
thiamin	25	vitamin B1	32
riboflavin	7	vitamin B2	14
pantothenic acid	14	vitamin B5	14

First, we tested the Consumer Preference score differences between using specialised and common utterances. We extracted all 46 health claims containing the specialised utterances (see Table 5), denoted as V_s. We developed another set, $V_{s \to c}$, by replacing the specialised utterances of health claims in V_s with the corresponding common utterances. Similarly, a set of 60 health claims containing the common utterances is denoted as V_c, while $V_{c \to s}$ consists of the health claims in which the common utterances in V_c are replaced with the corresponding specialised utterances. We computed the Consumer Preference scores for each health claim in V_s, $V_{s \to c}$, V_c and $V_{c \to s}$ respectively.

Next, we compared the Consumer Preference score differences between vitamins and minerals since minerals have no alternate name used in the food industry. Although vitamins and minerals belong to different nutrient categories, since vitamins are more common than mineral names in daily life, the latter is more obscure than the former for consumers. In line with the above method, we collected all 354 health claims containing vitamins, such as 'Vitamin A', 'B1' and 'C', developing a health claim set G_v. Its paired set $G_{v \to m}$ is developed by replacing a vitamin with a random mineral. Similarly, we gathered a mineral set G_m that has 191 health claims, and its corresponding vitamin set $G_{m \to v}$. We computed the Consumer Preference scores for G_v, $G_{v \to m}$, G_m and $G_{m \to v}$ respectively. The results are shown in Table 6a. By replacing specialised items with their corresponding common names, a large proportion of health claims (76%) achieved higher Consumer Preference scores. On the other hand, if common items were replaced with specialised names, most of the Consumer Preference scores (72%) decreased. This demonstrates that consumers prefer the common names of vitamins to their academic names in health claims. The comparison between vitamins and mineral shows that changing minerals to vitamins can yield higher Consumer Preference scores, while the reverse brings negative impacts. Thus, consumers prefer vitamins to minerals in health claims. This can be explained by the fact that consumers prefer common or familiar concepts in health claims on food in their daily life.

We further verify these statistical findings with human evaluation. We randomly select Health Claims from V_s, V_c, G_v, and G_m, respectively. Health Claims from V_s and V_c are paired for evaluating, when the survey is to test whether the common utterances are preferred to the specialised utterances. Similarly, health claims from G_v and G_m are paired to test whether vitamins are preferred to minerals. The survey was conducted based on the same method described in Sect. 3.1. We gathered 120 valid answers (6 tasks per person) from 20 subjects via Amazon MTurk. The statistical results are shown in Table 6b. Just as with the statistical findings, the human evaluation supports the argument that specialised utterances may reduce the attractiveness of health claims. Compared with conducting human evaluation, our automatic attractiveness analysis model is more efficient and simpler.

Table 6. Changes of Consumer Preference scores and human evaluation results

	CP score ↑		C P score ↓	
	count	ratio	count	ratio
$V_s \to V_{s \to c}$	35	**.76**	11	.24
$V_c \to V_{c \to s}$	17	.28	37	**.72**
$G_v \to G_{v \to m}$	103	.29	251	**.71**
$G_m \to G_{m \to v}$	163	**.85**	28	.15

(a) Changes of Consumer Preference scores by using alternative names (V) and different nutrients (G). The p-values are less than 0.0005 based on Mann-Whitney U test.

Hypothesis	Group	Ratio of chosen	p-value
$V_s \succ V_c$	V_s	.63	< .006
	V_c	.37	
$G_v \succ G_m$	G_v	.60	< .029
	G_m	.40	

(b) The human evaluation results to compare alternative names and different nutrients. The p-values are calculated via Mann-Whitney U test.

5.2   Sentiment and Metaphoricity Factors

This section investigates the influence of sentiment and metaphoricity features, which are the global features for sentences. For example, *"Natural source of vitamin A : contributing to boost your immune system"* has positive sentiment and contains a metaphor. Both sentiment and metaphoricity features are pragmatic features. health claims with different levels of sentiment polarities could emotionally impact consumer decisions. We computed the Pearson Correlation Coefficient between Consumer Preference scores and sentiment scores for all the health claims that we collected in Sect. 3.1. Sentiment scores are given by the model of [7] (here we only use the overall sentiment score). We visualise the distribution of the sentiment scores by Consumer Preference score in Fig. 6a. The moderate correlation coefficient (0.40 with p-value $< 10^{-5}$) suggests that health claims with positive sentiment are more attractive than negative Health Claims.

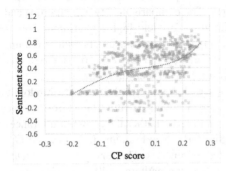
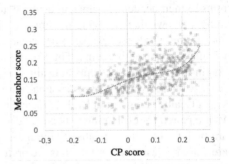

(a) Visualisation of the correlation for Consumer Preference score (CP score) and sentiment score among the health claim collection with an order 4 polynomial trendline. Their Pearson Correlation Coefficient is 0.40 with p-value $< 10^{-5}$.

(b) Visualisation of the correlation for Consumer Preference score (CP score) and metaphoricity score among the health claim collection with an order 4 polynomial trendline. Pearson correlation coefficient is 0.52 with the p-value $< 10^{-5}$.

Fig. 6. Sentiment and Metaphor

Metaphoricity is another pragmatic feature. Metaphorical expressions convey information conceptually [11], because they use one or several words to represent a different concept, instead of the original literal concepts. Thus, consumers may receive richer information from metaphoric health claims. Similar to sentiment features, we studied the correlation between Consumer Preference scores and metaphoricity scores. The metaphoricity scores are given by the model of [14], whose values are in a range of [0, 1]. The higher the score, the more likely the health claim is metaphoric. The correlation coefficient score (0.52 with p-value $< 10^{-5}$) suggests that consumers likely prefer health claims that use metaphoric expressions.

The above statistical analysis signifies that using words with positive sentiment polarities and metaphoric language in health claims can attract more consumers. We conducted a human evaluation to verify the hypothesis that health claims with higher sentiment/metaphor scores are more attractive. We paired the health claims with high sentiment scores (S_H with a sentiment score above 0.5) and health claims with low sentiment score (S_L with a sentiment score below -0.5). We also paired the high metaphoricity score health claims (S_H with a metaphoricity score above 0.2) with low metaphoricity score health claims (S_L with a metaphoricity score below 0.1). The human evaluation followed the same process as in Sect. 5.1. We recruited 20 valid subjects and gathered 400 answers from them. As seen in Table 7, the human evaluation results support our hypothesis. In practice, one may choose positive lexicons or use metaphoric expressions in writing health claims, which attract consumers.

Table 7. The human evaluation results of sentiment and metaphor factors. The p-values are calculated via Mann-Whitney U test.

Hypothesis	Group	Ratio of chosen	p-value
$S_H \succ S_L$	S_H	.58	$<.003$
	S_L	.42	
$M_H \succ M_L$	M_H	.55	$<.042$
	M_L	.45	

6   The Deployment of the Proposed Attractiveness Analysis Model

The proposed attractiveness analysis model is one major component of our funded project.[3][4][5]. The project had two components: a consumer toolkit and a manufacturer platform (called "Research, Analytics, and Consumer Insights Platform"). The consumer toolkit provides multiple interactive online *activities* including educational activities, and practice activities to teach health claim knowledge to consumers. This toolkit also tests and collects users' data about their understanding of the attractiveness of health claims. The manufacturer platform aims to support food manufacturers to evaluate their created health claims. The proposed attractiveness analysis model is used as the prediction engine of the manufacturer platform. By learning the consumer preferences from the collected data, there are two NLP-based prediction models in the manufacturer platform. The proposed attractiveness analysis model predicts how much consumers might like the Health Claims with 5-scale scores for two different scenarios. The first model predicts the general consumer preference score, which reflects the preference of the population. The second model is a conditional model, which predicts the target consumer preference scores – the preference of consumer characteristics (e.g. gender, age, etc.) is specified by the platform users. A detailed description of the design and implementation of the Consumer Toolkit is shown in our previous published system demonstration paper [12] (Figs. 7 and 8).

---

[3] Project website:https://www.healthclaimsunpacked.co.uk/.

[4] Consumer Toolkit website: https://www.unpackinghealthclaims.eu/.

[5] Manufacturer Platform website: https://www.healthclaimsinsights.eu/.

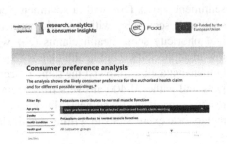

Fig. 7. The home page of the consumer toolkit

Fig. 8. The home page of the manufacturer platform

For the given health claims, the manufacturer platform can show the general consumer preference scores for the population, and the target consumer preference scores for groups of consumers by their characteristics (e.g. age-based groups, gender-based groups etc.). It suggests different wordings for the query health claim with the same nutrient and health benefit. The predicted consumer preference scores of the suggested health claims are shown to the users, which helps users to make decisions. A prototype of the consumer toolkit was released in English in November 2019. Versions of the consumer toolkit in five other European languages including German, French, Polish, Romanian and Hungarian were released in December 2021. Data from the consumer toolkit were used as a foundation to develop the manufacturer platform. The manufacturer platform was released in early 2022. It only has English version so far. It has the potential to be extended for predicting the attractiveness of health claims in other languages.

7   Conclusion

This paper discussed how to apply NLP techniques to better support the food industry for attractiveness analysis of health claimss. By introducing the new NLP task – Attractiveness Analysis, we developed a novel model to predict the consumer preferences of health claims. The model was trained on a newly collected health claim dataset with an improved Learn-to-rank loss function. By explicitly focusing on the syntactic and pragmatic features, the model successfully predicts consumer preference with high accuracy. Based on this model, we investigated and validated three important attractiveness factors. We observed that using common names instead of specialised academic names in health claims is more attractive. In addition, positive and metaphoric lexicons are also preferred. Our model can help manufactures evaluate their health claims without conducting a human-based survey. We also discussed the deployment of the proposed model in the manufacturer platform of the project system. In the future, we will explore a data-driven approach to identify more attractiveness factors and develop automatic attractiveness analysis tools for multi-lingual health claims. Also, we will consider the legal requirements and explore novel NLP tools to support food manufactures in designing health claims that are both attractive and legitimate (i.e., not deviating from the meaning of the original EFSA approved claim).

Acknowledgements. This project is funded by European Institute of Innovation and Technology (EIT) Food/EU Horizon 2020 (project number 19098). The authors would like to thank the funder EIT food and the great help and support of other team members of the project.

References

1. Saba, A., et al.: Country-wise differences in perception of health-related messages in cereal-based food products. Food Qual. Prefer. **21**(4), 385–393 (2010)
2. Bello, I., et al.: Revisiting resnets: improved training and scaling strategies. arXiv preprint arXiv:2103.07579 (2021)
3. Cambria, E., White, B.: Jumping NLP curves: a review of natural language processing research. IEEE Comput. Intell. Mag. **9**(2), 48–57 (2014)
4. Cao, Z., Qin, T., Liu, T.Y., Tsai, M.F., Li, H.: Learning to rank: from pairwise approach to listwise approach. In: ICML 2007, pp. 129–136 (2007)
5. Euromonitor: Functional foods - a world survey (2020)
6. Hussein, D.M.E.D.M.: A survey on sentiment analysis challenges. J. King Saud Univ.-Eng. Sci. **30**(4), 330–338 (2018)
7. Hutto, C., Gilbert, E.: Vader: a parsimonious rule-based model for sentiment analysis of social media text. In: Proceedings of the International AAAI Conference on Web and Social Media, vol. 8 (2014)
8. Jurafsky, D.: 26 pragmatics and computational linguistics. Handb. Pragmatics, 578 (2004)
9. Krishnamurthy, P., Carter, P., Blair, E.: Attribute framing and goal framing effects in health decisions. Organ. Behav. Hum. Decis. Processes **85**(2), 382–399 (2001)
10. Lähteenmäki, L.: Claiming health in food products. Food Qual. Prefer. **27**(2), 196–201 (2013). Ninth Pangborn Sensory Science Symposium
11. Lakoff, G., Johnson, M.: Metaphors We Live by. University of Chicago press (1980)
12. Li, X., Liang, H., Liu, Z.: Health claims unpacked: a toolkit to enhance the communication of health claims for food, pp. 4744–4748. Association for Computing Machinery, New York, NY, USA (2021). https://doi.org/10.1145/3459637.3481984
13. Li, Z., Cai, J., He, S., Zhao, H.: Seq2seq dependency parsing. In: ICCL 2018, pp. 3203–3214 (2018)
14. Mao, R., Li, X.: Bridging towers of multitask learning with a gating mechanism for aspect-based sentiment analysis and sequential metaphor identification. In: Proceedings of the 35th AAAI Conference on Artificial Intelligence (2021)
15. Mao, R., Lin, C., Guerin, F.: Word embedding and WordNet based metaphor identification and interpretation. In: ACL 2018, vol. 1, pp. 1222–1231 (2018)
16. Martínez, A., Mammola, S.: Specialized terminology reduces the number of citations of scientific papers. Proc. R. Soc. B **288**(1948), 20202581 (2021)
17. Roe, B., Levy, A.S., Derby, B.M.: The impact of health claims on consumer search and product evaluation outcomes: results from FDA experimental data. J. Publ. Policy Market. **18**(1), 89–105 (1999)
18. Bech-Larsen, T., Grunert, K.G: The perceived healthiness of functional foods: a conjoint study of Danish, Finnish and American consumers' perception of functional foods. Appetite **40**(1), 9–14 (2003)

19. Tversky, A., Kahneman, D.: The framing of decisions and the psychology of choice. Science **211**(4481), 453–458 (1981)
20. Van Kleef, E., van Trijp, H.C., Luning, P.: Functional foods: health claim-food product compatibility and the impact of health claim framing on consumer evaluation. Appetite **44**(3), 299–308 (2005)
21. Vaswani, A., et al.: Attention is all you need. In: NIPS (2017)

SchemaDB: A Dataset for Structures in Relational Data

Cody Christopher[1,3(✉)] , Kristen Moore[1,3] , and David Liebowitz[2,3]

[1] CSIRO Data61, Canberra, ACT, Australia
{cody.christopher,kristen.moore}@data61.csiro.au
[2] Penten Pty Ltd, Canberra, ACT, Australia
david.liebowitz@penten.com
[3] Cyber Security Cooperative Research Centre, Joondalup, WA, Australia

Abstract. In this paper we introduce the SchemaDB dataset; a collection of relational database schemas in both `sql` and graph formats. Databases are not commonly shared publicly for reasons of privacy and security, and so the corresponding schema for these databases are often not available for study. Consequently, an understanding of database structures in the wild is lacking, and most easily found examples of schema found publicly belong to common development frameworks or are derived from textbooks or engine benchmarks. SchemaDB contains 2,500 samples of relational schema found in public code repositories which have been standardised to `MySQL` syntax. We provide our gathering and transformation methodology, summary statistics, structural analysis, and discuss potential downstream research tasks in several domains.

Keywords: Web data collection · Data transformation · Datasets · Relational databases · Machine learning

1 Introduction

An on-going problem in the machine learning research community is a shortage of suitable datasets. Often, the release of data is impeded by concerns surrounding intellectual property, disclosure, or privacy. This hampers efforts to replicate, extend, and compare results across a research domain [10].

One example where publicly available data is lacking is that of database schemas. The majority of databases available publicly are designed for benchmark evaluation of the performance of various SQL engines, or developed for education purposes to demonstrate design principles. Unfortunately, these are not suitable and varied enough representations from which to learn the structure of real world databases, such as those one might expect to find inside a corporate network.

This work has been supported by the Cyber Security Research Centre Limited whose activities are partially funded by the Australian Government's Cooperative Research Centres Programme.

A rich and standardised dataset of diverse schema would be useful for developing models for database analysis and simulation. Such analysis could include the study of common database structures, the prevalence of best practice implementation strategy (e.g. degree of normalisation), and common database purposes and types. A database dataset would also unlock the potential to train models for database augmentation, generation, and simulation. A database generation model could be used for example in the cyber security domain to create realistic content for cyber ranges or for honeypots to be used for cyber deception. Bringing automation and scale to the generation of diverse and realistic content is highly beneficial in these fields, where the typical practice of hand-crafting of realistic content is very costly [8].

To close this gap, we introduce SCHEMADB[1], a collection of 2,500 standardised schemas collected from real projects, largely found on GitHub. We discuss our gathering and transformation methodology, and provide an analysis of the structure and summary statistics of the collection. We release the curated dataset (and extensions) alongside the paper, as well as the code used to extract schema from a collection of repositories such that the dataset can be replicated or further expanded.

1.1 Existing Datasets

To the best of our knowledge there is only one other public dataset that has similar purpose and viable schema – the CTU Prague Relational Learning Repository [9]. The stated purpose of this dataset is to enable machine learning on relational data, as opposed to single table (the majority of data releases). It includes full databases (inclusive of data) on a public database server for this purpose. After excluding benchmarking or sample databases, there are 62 samples. We incorporate these into SCHEMADB as the only samples not from GitHub. The primary advantage of the CTU dataset is that full sample data can be obtained to enable machine learning tasks that rely on this.

1.2 Challenges of Flat Data

Well-known dataset collections such as Kaggle or Zenodo are useful sources of large datasets, often prepared for machine learning tasks. In theory, these collections could provide suitable examples of relational datasets, but the data is typically available in flat table format.

The Normal Form (*NF) hierarchy of database structures describe the various levels to which a dataset might be normalised to allow for easy querying, minimisation of data redundancy, and ease of maintenance [5]. In the process of designing a database, often an architect will start with a collection of the entities and attributes to be stored, and subsequently decide how these entities relate to each other. The decisions made here, as to what each table in a schema

---

[1] Downloadable at https://bitbucket.csiro.au/projects/DECAAS/repos/schema_db.

represents and how they relate, are commonly referred to as normalisation. Commonly described in textbooks on database design, it is suggested that the third form (3NF) or Boyce-Codd [4] form (BCNF, sometimes called 3.5NF) is the first form at which a database is suitable for implementation.

To reach 3NF from a flat dataset involves the identification of the structural rules which govern the data, known as Functional Dependencies (FD). From a collection of FDs, algorithms exist which can automate the normalisation to 3NF (or higher), however, the process of identifying the FDs themselves is still one requiring human level understanding of the semantics of the data. There is some limited research towards the automatic identification of FDs from data using machine learning techniques [14], however this is not mature enough that it would be possible to automate the collection of schemas from flat file dataset repositories. Consequently, we have not used these data sources in the construction of SCHEMADB.

2   Dataset Curation

In this section we detail the process of collecting and curating SCHEMADB, providing context for various data collection issues and our proposed solutions.

2.1   Collection and Filtration

The primary data source for SCHEMADB is a collection of open source repositories found on GitHub. Sources were selected by searching for repositories with a telltale `schema.sql` file containing a `CREATE DATABASE` statement. There are a number of challenges faced here, including:

- Repositories for large projects will often contain many scripts identifiable as 'schema', however, these can often take the form of migration scripts (version updates), architecture targets (the same schema in multiple dialects), adjusting permissions, or data insertion. Capturing all of these would lead to considerable duplication.
- Large projects may also include the codebase of common libraries they rely on, which can cause the appearance of multiple disparate schema per repository.
- Varying `.sql` dialects support different levels of referential integrity, user-specified rules, and hook-in scripts (i.e. those run before or after a certain operation).
- The dialect of `.sql` used in a schema is often unspecified, and a mismatch can cause parsing failure. Substantial differences between versions of the main dialects can also result in failure to parse.

The net effect of these challenges means that, without manual inspection, repositories that match these edge cases introduce substantial overhead to an automated collection process. We have developed heuristics to capture a large proportion of the schemas.

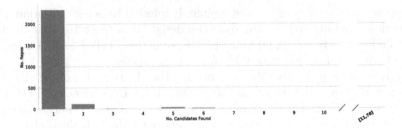

Fig. 1. Histogram of found schema candidates per repository.

To avoid unnecessary manual filtration of thousands of files within thousands of repositories, only the largest schema (in terms of number of tables described) from each repository was captured as a representative sample, assuming that a larger schema is more likely to have some degree of normalisation applied for ease of management. This has the effect of introducing some edge cases where files such as migration scripts (as opposed to the underlying schema) are captured from repositories that have archived migrations. An analysis of these outliers indicates that these are rare and often still represent valid (sub-)schema. Figure 1 shows that the vast majority of repositories only had a single a candidate, and inspection of those that had 2 or more revealed they were often duplicates or migrations.

We consider only those schemas that can be parsed with common encodings (utf8 and utf16) as this covers the vast majority of conforming schema. It may be worth extending this in the future to capture schema in less common encodings and capture those with names written in languages requiring extended symbols. As the primary purpose of this collection is to provide a basis for investigating relational structures, we filter initially to those schema that contain at least two tables such that relations can actually be defined. Edge cases were found in the raw data when commented code was detected indicating the presence of user intervention (as opposed to database engine generation as the result of a schema dump, or other tool assistance). For candidates that meet the prior conditions, this is subsequently handled by the compiled parsers.

2.2   Graph Transform and Canonisation

To standardise the data format for subsequent analysis, we modified the existing grammars and constructed custom parsers for the three most common SQL dialects; MySQL, SQLite, and postgreSQL. As the most permissive of the dialects by design, MySQL was selected as the output format for canonisation. To assist with graph analytics, a directed heterograph representation was generated from the initial parse, where necessary alterations could be performed such that valid SQL could be generated in the target dialect. This included performing a data type mapping as part of the transpilation to account for the major classes of attribute typing across dialects.

We propose that in an analysis of relational structure, modern techniques in graph analysis and graph neural networks may be useful in uncovering latent structural properties and similarities that otherwise would go unnoticed. As the respective parse-trees of each dialect allow for easy conversion to and from a graph-based format, we take this opportunity to use a graph as an interchange medium. Subsequently, we are also able to provide these graphs in a common format alongside the standardised schema.

Graphs are constructed by creating nodes for each table, column, and foreign key linkage, where nodes are explicitly typed. Edges in these graphs are not explicitly typed but can be inferred from the direction and the connecting nodes. Edges outbound from a table node to column node indicate that the column belongs to the table. Edges outbound from a column to a foreign key node indicate that this is a *referencing* column, and edges outbound from a foreign key to a column indicate a *referenced* column. Foreign key nodes can have multiple inbound or outbound edges in the case of compound foreign keys (involving multiple columns), but as a collection the columns on either side will always be incident to the same source and destination tables, and the inbound and outbound degrees will be identical. Nodes also contain the semantic information associated with their type. For tables this information includes the table name and primary key(s), for columns this is their name, data type, optional data length, and position, and for foreign keys this is the name of the constraint (if explicit) and ordered lists of the columns in the key as this cannot be derived from the edges. We illustrate this graph view in a simple schema in Fig. 2.

For a higher level dependency analysis, what we call a *skeleton* of this graph is provided. The skeleton is a regular graph, where nodes denote tables and directed edges are foreign key dependencies. The edge direction here is reversed, with an outbound edge indicating that the destination *references* the source.

Additional parsing rules were utilised as many dialects support declaring key constraints separately to table creation, and some methods of automated schema dumps perform this automatically. Some dialects also often support designing multiple sub-schemas in the same database. For ease of analysis, we count these as single database entities and adjust duplicate table names as necessary. Some manual filtration of outliers at the extreme ends of the distribution (large amount of tables, or single tables) was carried out to remove schemas designed for benchmarking, and these are not present in the released data.

At this stage each graph skeleton is then topologically sorted and this order is used for canonisation to valid MySQL, with obfuscation of the database name such that individual source repositories are not easily identifiable. Topological sorting ensures that the tables are created in an order conducive to foreign key enforcement (the referenced column exists prior to the referencing column), although this is not strictly necessary when the entire database is created in a single transaction.

For a final consistency check, we make use of existing techniques to check for duplication in the corpus. Existing software approaches designed for plagiarism analysis are suitable here as they support detection of structure even in the

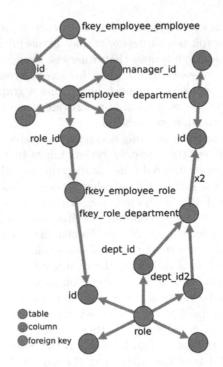

Fig. 2. Graph representation of a relational schema. Some nodes left unlabelled for brevity.

presence of changing variable names. Specifically, we make use of the Measure of Software Similarity (MOSS) [11] as the primary duplication detection method. When a high degree of similarity is detected, only one sample amongst all identified duplicates is kept for the canonical set. However we provide the degree of duplication and the excluded samples separately, should they be deemed necessary. Roughly 30 samples were removed as a result of this analysis.

2.3   Heuristic Augmentation

After the initial canonisation pass described in the previous section, further analysis and curation is possible. For example, during analysis it was discovered that some samples containing more than two tables had no foreign keys (and thus no explicit relational structure). Whilst it would be ideal to perform a normalisation analysis, there is presently no way to do this without performing an analysis of functional dependencies on every sample manually, which we consider a downstream task if desirable at all. It is, however, likely that samples like this in the collection were not normalised according to standard practice.

Missing Foreign Key Imputation. One avenue of augmentation is to try and find implied (missing) foreign keys. As a result of common naming conventions, we demonstrate that it is possible to make explicit these implicit keys. It is common practice is to have an ID column as the primary key of entity, which we denote entityA for the purposes of the example. If there is another table with references to this entity, the keyed column will typically directly reference this primary key and subsequently the column name will reflect this: entityA_ID or entityAID or similar.

We show the results of performing this type of analysis under a variety of heuristics. First, we consider only direct matches – where a column name is exactly the underscore concatenation of another table name with a column in that table. We then allow a progressive increase of the edit (Levenshtein) distance to the generated identifiers to account for common variations, acronyms and idioms. We observe than beyond a distance as small as three, the frequency of spurious matches becomes untenable due to prevalence of short column names and abbreviations. We provide in the release the exact (or distance zero) matches separately, and include the script to generate foreign keys for greater distances in the repository.

3   Analytics

3.1   Summary Statistics

We present a breakdown of the summary statistics of the schema present in the SCHEMADB. Note that the histograms are truncated due to very long tails. The distribution in the tails is shown with varying bin sizes to give an indication of the density and length of each tail.

In Fig. 3 we present a breakdown of the data in terms of the size of each database with respect to the number of tables present. In Fig. 4 we further breakdown the size of tables with respect to the number of columns present. In Fig. 5 we show the prevalence of foreign keys as the number of these found in each database, excluding the zero cases. In Fig. 6 we show, the number of recovered foreign keys recovered in roughly 500 of the cases where no foreign key was recorded.

Note the recovered foreign keys with exact matching according to our naïve heuristic, as keys were recovered in approximately 20% of all samples lacking foreign keys, indicating that keys are frequently present but often unspecified.

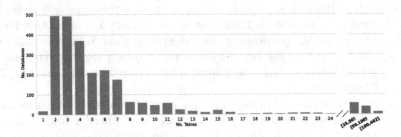

Fig. 3. Database size statistics

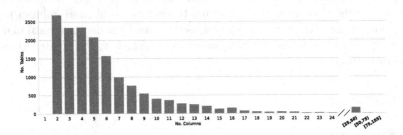

Fig. 4. Table size statistics

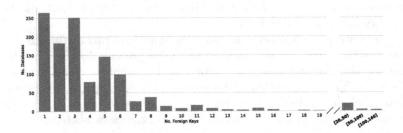

Fig. 5. Foreign key count statistics

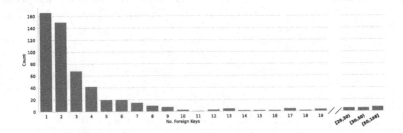

Fig. 6. Recovered foreign key count statistics

4   Research Potential and Applications

We present several possible downstream investigations enabled by the existence of this data set. To the best of our knowledge there are no existing analyses of relational schema in the wild. From a research perspective there are several interesting avenues worth pursuing.

With respect to how databases are used in practice, we pose the following questions:

1. How often are relational databases normalised (and to what degree) in practice?
2. How often do specific entities appear and with what frequency do the co-occur?
3. By what names do common entities go by (i.e. person, user, customer, agent, employee) and can broad database purpose be determined from semantics?

The first of these is of particular interest, as there is no way to determine which normal form a database is in without performing a manual analysis. A classifier that could determine the normal form of a schema would help in the development of optimally normalised databases. This would require labelling the schema for a supervised approach.

In multi-entity databases, some contextual signs of bad design are often evidenced by the following:

- A single monolithic table[2]
- Multiple entity types existing in the same table
- Absence of foreign keys linking clearly dependant columns across entities

We suggest that a database assistance agent that can identify normalisation status and detect entities could assist in the elimination of these types of issues.

In respect of the remaining questions (2 and 3) above, classification approaches in AI and ML could be considered. In particular the use of language models such as GPT [2] with fine-tuning on novel domains (e.g. Image-GPT [3]) could assist with these determinations by way of clustering in the embedding space.

This leads us to the question of automated generation, where there are a number of practical applications to consider. In particular, using both language models (e.g., GPT family) and novel heterogeneous graph models, we see as several feasible application domains:

1. Is it possible to generate relational schema automatically from input text – such as a requirements specification?
2. Can existing data generation approaches be used in conjunction with schema generation to generate entirely novel databases with minimal to no prompting?

---

[2] with regards to databases in relational engines. This convention is typically intentionally not followed in large flat-file highly distributed engines such a BigTable, Hadoop, etc.

3. Can this generation process be tuned for the creation of assets intended for cyber security, in particular cyber deception?

Synthesising databases has application in cyber deception, where databases can be used as honeypots, or database elements as honeytokens [1,12]. Additionally, cyber research making use of realistic environments (such as reinforcement learning, or cyber ranges) could also benefit from the ability to use generated databases to increase (or decrease) the perceived realism of the environment. In terms of generative models, we also envision that the provided graph representations will enable approaches similar to that used in cutting edge arbitrary graph generation [6,7,13,15].

5   Conclusion

The SCHEMADB dataset of relational database schemas supports the study and development of a variety of machine learning applications, as well providing an initial standardised example for other potential dataset builders. A clear limitation of SCHEMADB is that it is restricted to freely available, public datasets, and the types of schemas accessible on GitHub may not be representative of the database population on the internet, within large corporations, and in proprietary commercial applications. We therefore hope that the release of SCHEMADB will encourage others to release similar datasets to augment this initial corpus. We also see the potential to grow SchemaDB by automating the data collection and ETL scripts periodically, and by incorporating other code repository hosts such as GitLab. We further intend to provide an API and interactive site for exploring the dataset subsequent to release.

References

1. Abay, N.C., Akcora, C.G., Zhou, Y., Kantarcioglu, M., Thuraisingham, B.: Using deep learning to generate relational HoneyData. In: Al-Shaer, E., Wei, J., Hamlen, K.W., Wang, C. (eds.) Autonomous Cyber Deception, pp. 3–19. Springer, Cham (2019). https://doi.org/10.1007/978-3-030-02110-8_1
2. Brown, T.B., et al.: Language models are few-shot learners. In: Larochelle, H., Ranzato, M., Hadsell, R., Balcan, M.F., Lin, H. (eds.) Advances in Neural Information Processing Systems (NeurIPS 2020), vol. 33, pp. 1877–1901. Curran Associates, Inc. (2020). https://proceedings.neurips.cc/paper/2020/file/1457c0d6bfcb4967418bfb8ac142f64a-Paper.pdf. https://papers.nips.cc/paper/2020
3. Chen, M., Radford, A., Child, R., Wu, J., Jun, H., Luan, D., Sutskever, I.: Generative pretraining from pixels. In: International Conference on Machine Learning, pp. 1691–1703. PMLR (2020)
4. Codd, E.: Recent investigations into relational data base. In: Proceedings 1974 IFIP Congress (1974)
5. Elmasri, R., Navathe, S.B., Elmasri, R., Navathe, S.: Fundamentals of Database Systems. Springer, Cham (2000)

6. Li, Y., Vinyals, O., Dyer, C., Pascanu, R., Battaglia, P.: Learning deep generative models of graphs. arXiv preprint arXiv:1803.03324 (2018)
7. Liao, R., et al.: Efficient graph generation with graph recurrent attention networks. In: Advances in Neural Information Processing Systems, pp. 4257–4267 (2019)
8. Liebowitz, D., et al.: Deception for cyber defence: challenges and opportunities. In: 2021 Third IEEE International Conference on Trust, Privacy and Security in Intelligent Systems and Applications (TPS-ISA), pp. 173–182. IEEE (2021)
9. Motl, J., Schulte, O.: The ctu prague relational learning repository. arXiv preprint arXiv:1511.03086 (2015)
10. Paullada, A., Raji, I.D., Bender, E.M., Denton, E., Hanna, A.: Data and its (dis) contents: a survey of dataset development and use in machine learning research. Patterns **2**(11), 100336 (2021)
11. Schleimer, S., Wilkerson, D.S., Aiken, A.: Winnowing: local algorithms for document fingerprinting. In: Proceedings of the 2003 ACM SIGMOD International Conference on Management of Data, pp. 76–85 (2003)
12. Spitzner, L.: Honeypots: Catching the insider threat. In: 19th Annual Computer Security Applications Conference, Proceedings, pp. 170–179. IEEE (2003)
13. Stier, J., Granitzer, M.: DeepGG: a deep graph generator. In: Abreu, P.H., Rodrigues, P.P., Fernández, A., Gama, J. (eds.) IDA 2021. LNCS, vol. 12695, pp. 313–324. Springer, Cham (2021). https://doi.org/10.1007/978-3-030-74251-5_25
14. Tahvili, S., Hatvani, L., Felderer, M., Afzal, W., Bohlin, M.: Automated functional dependency detection between test cases using doc2vec and clustering. In: 2019 IEEE International Conference On Artificial Intelligence Testing (AITest), pp. 19–26. IEEE (2019)
15. You, J., Ying, R., Ren, X., Hamilton, W.L., Leskovec, J.: GraphRNN: generating realistic graphs with deep auto-regressive models. In: Proceedings of the 35th International Conference on Machine Learning, PMLR, vol. 80, pp. 5708–5717 (2018). https://proceedings.mlr.press/v80/

Author Index

Austin, Eric   148

Bashar, Md Abul   133
Brach, Paweł   205

Catchpoole, Daniel R.   58
Chen, Bowen   28
Christopher, Cody   233

Dobrakowski, Adam Gabriel   205

Esmaili, Nazanin   15
Estivill-Castro, Vladimir   115

Fitzpatrick, Stuart   3
French, Tim   176

Gilmore, Eugene   115

Halstead, Ben   28
Hexel, René   115
Hodkiewicz, Melinda   176
Huynh, Du   43

Islam, Md Zahidul   73

Jones, Rodney   217

Kennedy, Paul J.   58
Koh, Yun Sing   28, 90
Kumar, Amit   15

Largeron, Christine   148
Li, Bing   99
Li, Xiao   217
Li, Yuefeng   163
Liang, Huizhi   217
Liebowitz, David   233
Liu, Wei   43, 176
Liu, Zehao   217

Mammadov, Musa   192
Moore, Kristen   233
Mucha, Marcin   205
Muspratt, Rob   192

Naqvi, Naureen   73
Nayak, Richi   133
Nguyen, Quang Vinh   58

Obst, Oliver   3

Pacuk, Andrzej   205
Park, Laurence   3
Piccardi, Massimo   15
Pingi, Sharon Torao   133

Qu, Zhonglin   58

Ramesh, Krithik   90
Rehman, Sabih Ur   73
Reynolds, Mark   43
Ryder, Chris   217

Sankowski, Piotr   205
Simoff, Simeon J.   58
Stewart, Michael   176
Sun, Qiang   43

Tai, Han   99
Tegegne, Yezihalem   58
Trabelsi, Amine   148

Wijesinghe, Yashodhya V.   163
Wong, Raymond   99

Xu, Yue   163

Zaïane, Osmar R.   148
Zhang, Qing   163
Zhao, Ziyu   176

Printed in the United States
by Baker & Taylor Publisher Services